D0987037

CLEANING UP COAL:

A Study of Coal Cleaning and the Use of Cleaned Coal

CLEANING UP COAL:

A Study of Coal Cleaning and the Use of Cleaned Coal

authors
Cynthia A. Hutton
Robert N. Gould

research consultant
Sophie R. Weber

editor
Richard C. Allen

an INFORM book

Ballinger Publishing Company
Cambridge, Massachusetts
A Subsidiary of Harper & Row, Publishers, Inc.

INFORM, Inc.
381 Park Avenue South
New York, N.Y. 10016
212/689-4040

Copyright © 1982 by INFORM, Inc.
All rights reserved.

ISBN 0-918780-18-7

Library of Congress #82-81548

Cover design by Philip A. Scheuer

Cover typography by Chance

This report was prepared wholly under contract with the United States Government. Neither the United States nor the United States Department of Energy nor any of their employees, nor any of their contractors, subcontractors, or their employees, makes any warranty, express or implied, or assumes any legal liability or responsibility for the accuracy, or completeness, of any information, apparatus, product, or process disclosed, or represents that its use would not infringe privately owned rights.

The views and conclusions contained in this report should not be interpreted as representing the official policies, either expressed or implied of the U.S. Department of Energy.

CLEANING UP COAL

Table of Contents

Appendix

Preface

In this report, INFORM provides facts that will help the non-technical decisionmakers in the U.S. understand a technology that can significantly reduce the polluting effects of burning coal. Those decisionmakers include legislators, regulators and utility executives, public interest groups, concerned community organizations and environmentalists who, for much of the last decade, have been involved in the debate over the merits of the broader use of our most abundant fossil fuel–coal. The use of this resource, especially in large industrial and utility plants, has created widespread and intense public controversy.

For the past four years INFORM has turned its research capabilities to defining cleaner and more economical ways of using U.S. coal supplies. In a series of studies we have explored the state of coal gasification and liquefaction alternatives; we have assessed better ways to reclaim strip-mined coal lands in the West; we have studied the potential for using a technique called fluidized-bed combustion for cleaner burning of coal in industrial and utility operations. And in this study, and in another about to be issued, we have focused on finding out what cleaning coal and using flue gas desulfurization systems (called "scrubbers") can contribute to reducing the polluting effects of burning coal in utility plants.

In all its research, INFORM takes the case study approach. We document the current effects of industrial practices and explore options for improved environmental performance. We assess current benefits and costs of using various pollution controls and resource conservation programs. We explore the problems producers and users of such techniques have experienced, and their views on future markets.

All in all, both scrubbers and coal cleaning offer exciting and important possibilities for putting more coal to work in generating power in this country more economically and still meeting critical air quality standards that have been set to protect public health. The need for accurate and clear information about these two technologies is evident: 80% of the sulfur dioxide emissions in the U.S. now come from utility power plant operations, and over 140 existing oil-fired power plants are candidates for conversion to coal use.

We earnestly hope that this documentation of the technologies for cleaning coal, along with INFORM's companion study of scrubber systems, may help government and business planners and concerned citizens chart intelligent future courses and set realistic goals for meeting our energy needs in an environmentally sound manner.

Joanna D. Underwood
Executive Director
INFORM

Acknowledgements

The authors would like to express their gratitude to those people without whose help this study would not have been possible. Sophie Weber, INFORM's Research Consultant, advised on and directed the project; we would also like to thank Robbin Blaine and Perrin Stryker, whose editorial guidance helped us immensely throughout the research and writing periods.

Jonathan Kalb, a student intern who participated in early research with great zeal and a special sense of humor, has our gratitude for his contribution — most notably his thorough investigations of chemical coal cleaning.

While INFORM accepts full responsibility for the content and conclusions of this study, much valuable outside advice and criticism has helped shape it. Portions of *Cleaning Up Coal* have been extensively reviewed by experts in various fields. We are grateful for the time and effort of: Richard Ayres, Natural Resources Defense Council; Randy Cole, Supervisor, Fuels Group, Tennessee Valley Authority; Robert L. Frank, Projects Manager, Coal Cleaning Performance and Reliability, Tennessee Valley Authority; Keith Frye, Acting Deputy Director, Office of Coal Processing, Fossil Energy, U.S. Department of Energy; James Kilgroe, Coal Program Manager, U.S. Environmental Protection Agency; Harry Perry, Consultant, Resources for the Future; George Sall, Senior Advisor, Division of Planning and Environment,

Fossil Energy, U.S. Department of Energy; and Randhir Sehgal, Project Manager, Coal Quality, Electric Power Research Institute.

Within INFORM, James Cannon provided invaluable technical assistance. Mary Ann Baviello was especially helpful, contributing her knowledge and inspiring good humor. Clifford Bob, Alexandra Bowie and Lydia Maruszko gathered information and contributed ideas during the course of the research. Richard Allen's editing helped to shape and clarify the study. Mary Maud Ferguson's excellent copyediting has created a report that is both readable as well as informative. Student interns Lisa Rosenfield, William Tuthill and Barry Wasserman contributed essential research and fact-checking over these past months. Ilene Green was most helpful in thoroughly researching the current status of the Clean Air Act. We'd also like to thank two people without whom there would have been only a mass of loose typed sheets and no footnotes: Risa Gerson and Susan Jakoplic, who with their patience, skill and sound judgment, put the book together.

Special thanks are also due to Viviane Arzoumanian, INFORM's Administrator who kept the office running, and to Patricia Holmes, Charles Lowy and Rose Valenstein for their excellent typing skills and great patience. We'd also like to express our appreciation to INFORM's Executive Director, Joanna Underwood and to the entire INFORM staff for their support.

CLEANING UP COAL:

A Study of Coal Cleaning and the Use of Cleaned Coal

Part I

Introduction

Coal cleaning is a flexible technology that, in its most primitive form, has been used almost as long as coal has been mined. Over recent decades, more advanced forms have continued to evolve. Basically, all cleaning processes aim to improve the quality of raw coal by removing varying amounts of sulfur, ash and other wastes before combustion. The resulting product is a higher quality fuel for utility and industrial boilers, and is easier on plant equipment.

INFORM has examined both the state of the art, and the state of practice of the coal cleaning processes currently being applied and developed. INFORM has also examined the impact of burning cleaned coal on one of the principal users of this fuel, electric utilities. In conducting this research, INFORM surveyed a broad range of companies that produce and use cleaned coal with the aim of defining the current costs and benefits associated with this technology, and of exploring the contribution it may make to cleaner and more efficient coal use in the future.

There are three basic kinds of cleaning, each with a wide range of capabilities. This report examined all three. Conventional physical methods, the most widely known and used to date, reduce the size of coal pieces and separate the coal from its impurities by exploiting the differences in specific gravity and/or surface properties between coal and waste minerals.

INFORM found in its sample of 12 preparation plants that an average of 25 percent of a coal's sulfur and 51 percent of a coal's ash was removed. The advanced physical methods attempt to clean coal more efficiently and at a lower cost than can conventional methods. Chemical coal cleaning, which can be viewed as the "true state of the art," breaks down the coal's molecular structure. It seeks to remove almost 100 percent of the inorganic sulfur and up to 90 percent of the total sulfur, while the most sophisticated conventional methods have, to date, removed only up to 47 percent of the total sulfur and none of the organic sulfur.

As neither chemical or the advanced physical processes are in commercial use, the main focus of INFORM's study is on the state of practice of conventional coal cleaning; however, research and development of advanced physical and chemical systems are also described.

Table A:

COAL PRODUCTION AND EXPORT

BITUMINOUS COAL, LIGNITE, AND ANTHRACITE

	Production	Exports
	(Thousand Short Tons)	
1973 Total	598,568	53,587
1974 Total	610,023	60,661
1975 Total	654,641	66,309
1976 Total	684,913	60,021
1977 Total	697,205	54,312
1978 Total	670,164	40,714
1979 Total	781,134	66,042
1980 Total	829,700	91,742
1981 (Thru May)	289,180	36,633

Source: Department of Energy, *Monthly Energy Review*, August 1981

The Expanding Use of Coal in the United States

Coal is abundant in the United States, accounting for about 82 percent of the nation's recoverable energy reserves.[1] Most analysts expect it to play a significant role in the world's future energy supply. Since the "energy crisis" of 1974, coal production in this country has steadily increased from 599 million tons in 1973 to 830 million tons in 1980.[2] The U.S. Department of Energy projects that by 1980, 1.2 billion tons will be produced.[3]

In 1947, utilities burned 86 million tons, or 16 percent of the bituminous and lignite coal consumed.[4] Their share has steadily increased. In 1980, they burned 569,274,000 tons, or 81 percent of the coal consumed in the United States (see Table B). By 1985, according

Table B:

COAL CONSUMPTION-

BITUMINOUS COAL, LIGNITE, AND ANTHRACITE

| | Electric Utilities | Industrial | | | Total |
		Coke Plants	Other Industrial Including Transportation	Residen- tial and Commer- cial	
	(Thousand Short Tons)				
1973 Total	389,212	94,101	68,154	11,117	562,584
1974 Total	391,811	90,191	64,983	11,417	558,402
1975 Total	405,962	83,598	63,670	9,410	562,641
1976 Total	448,371	84,704	61,799	8,916	603,790
1977 Total	477,126	77,739	61,472	8,954	625,291
1978 Total	481,235	71,394	63,085	9,511	625,225
1979 Total	527,051	77,368	67,717	8,388	680,524
1980 Total	569,274	66,660	60,347	6,452	702,733
1981 (Thru March)	150,669	17,735	17,997	1,890	188,290

Source: Department of Energy, *Monthly Energy Review*, August 1981

to predictions by the President's Commission on Coal, utility use will rise to 677,082,000 tons.[5]

Utility coal use plays an integral role in the U.S. coal market, and coal is by far the primary energy source for the production of electricity. In 1980, about 51 percent of the electricity generated in the United States was produced by burning coal—with natural gas a distant second at 15 percent[6] (see Table C).

Table C:

ELECTRIC UTILITIES

NET ELECTRICITY PRODUCTION BY PRIMARY ENERGY SOURCE

	Coal	Petroleum	Natural Gas	Nuclear	Hydro	Other	Total
			(Million Kilowatt-hours)				
1973 Total	847,651	314,343	340,858	83,479	272,083	2,294	1,860,710
1974 Total	828,433	300,931	320,065	113,976	301,032	2,703	1,867,140
1975 Total	852,786	289,095	299,778	172,505	300,047	3,437	1,917,649
1976 Total	944,391	319,988	294,624	191,104	283,707	3,883	2,037,696
1977 Total	985,219	358,179	305,505	250,883	220,475	4,063	2,124,323
1978 Total	975,742	365,060	305,391	276,403	280,419	3,315	2,206,331
1979 Total	1,075,037	303,525	329,485	255,155	279,783	4,387	2,247,372
1980 Total	1,161,562	245,994	346,240	251,116	276,021	5,506	2,286,439
1981 (Thru May)	485,333	89,744	125,549	107,336	108,865	2,547	920,375

Source: Department of Energy, *Monthly Energy Review*, August 1981

Increasing coal use, however, creates some potentially serious environmental problems. Compared with the other fossil fuels, oil and natural gas, coal is a dirty fuel. In its raw form, coal contains many impurities, such as rocks that line coal seams, organic and inorganic sulfur, ash, trace minerals, and other wastes. When burned, many of these impurities can be transformed into sulfur dioxide, particulates and other air pollutants.

In 1977, the latest date for which figures are available, power plants emitted 19.4 million tons of sulfur oxides.[7] These emissions contribute to an air pollution load that has been linked to respiratory and heart problems, and increasingly to the formation of acid precipitation, which is affecting hundreds of rivers and lakes in the United States and Canada, destroying fish and other aquatic life. A report released in October 1981 by the National Academy of Sciences concludes: "Although claims have been made that the direct evidence linking power plant emissions to the production of acid rain is inconclusive, we find the circumstantial evidence for their role overwhelming." According to this report, the acid rain phenomenon "is disturbing enough to merit prompt tightening of restrictions on atmospheric emissions of fossil fuels and other large sources."[8]

The level of reduction in sulfur oxide emissions to the atmosphere, and the technologies used to control these emissions, are currently the subjects of a national debate on the clean air laws in this country. The participants in this debate have taken a variety of sides on these issues.

Congress has recognized the need to significantly reduce sulfur oxide emissions. However, there is serious debate over what specific legislative action should be taken, particularly what changes, if any, should be made in the Clean Air Act. The Reagan Administration has not yet proposed a program for revising the Clean Air Act. Environmentalists want to expand flue gas desulfurization requirements to include older power plant boilers which are currently exempt from the Clean Air Act, and which release a major portion of the sulfur oxide emissions in the United States. Industry is resisting requirements to retrofit scrubbers on older plants, as well as any move at present to include new emission requirements in the Clean Air Act. According to many industry and government sources, it seems likely that industry will take the initiative to prevent moves to expand the use of scrubbers by offering to clean all coal, or a large portion of it.

Regardless of the outcome of this debate or whether any new scientific data linking acid precipitation to sulfur oxide emissions is produced, it is clear that if the nation is to rely on its coal reserves as a source of energy, identifying the technological options for using coal more cleanly—of which coal cleaning is one—is a high public priority. In fact, recent opinion polls reflect broad public support for protecting the air quality in this country.

The Evolving Role of Coal Cleaning

Originally, since the 1880's, coal cleaning was used almost exclusively by the metallurgical industry for coke manufacturing.[9] This process required a coal of uniform consistency with a low percentage of sulfur and ash. Utilities have become increasingly heavy users of cleaned coal for steam generation. In 1979, about 20 percent of the 527,051,000 tons of coal burned by utilities was cleaned to remove sulfur and ash impurities.[10]*

Cleaning offers a number of potential advantages to utilities, all of which INFORM has sought to define in this report. It is one of the several alternatives that can help utilities meet sulfur dioxide standards established by the Clean Air Act and its amendments. In addition to removing a portion of the ash and mining wastes, cleaning can provide a coal of improved quality, higher heating value and greater consistency. Although it is more expensive than it would be raw, cleaned coal may save utilities money by lowering their boiler maintenance and other plant operating costs, reducing the costs of transporting coal from the production site and increasing boiler efficiency.

Cleaning also offers the possibility of expanding the amount of coal appropriate for burning by converting a coal of low quality into one of higher quality. In the past four decades, the quality of coal being produced has steadily declined. One reason for this is that since the 1940s automatic mining has replaced mechanical mining and loading methods which allowed coal to be selected more carefully. A second reason is that many higher quality seams have been depleted, especially in the heavily mined Appalachian and midwestern regions.[12] As a result, greater amounts of waste are present in the raw coal, especially from coal from underground mines.

Finally, cleaning may reduce or eliminate flue gas desulfurization, or scrubber, requirements for utilities (depending on the sulfur content of the coal used, and local emission standards). If cleaning enables utilities to install smaller scrubbers to meet air quality standards, savings in capital, in operation and maintenance, and in waste handling costs should all be realized.

---

*Actually, about 70 percent of all coal used by utilities is cleaned to "minimal levels," which usually involves only crushing and breaking.[11]

Government Incentives

Government budgetary and regulatory policies have played an important, sometimes contradictory, role in the development and use of the coal cleaning processes. In general, the 1970 Clean Air Act has encouraged government funding for research and development of more advanced cleaning technologies and has led to the wider commercial application of higher levels of conventional coal cleaning. On the other hand, the 1977 Revised New Source Performance Standards have all but eliminated research and development support for chemical processes which do not have the potential to meet the 90 percent sulfur reduction standard which these standards require. The impact of these regulations on the commercial use of coal cleaning appears to have been mixed, depending on a variety of factors related to the particular standards to be met by individual plants at specific locations.

Government funded research and development of coal cleaning technologies began in the 1940s and 1950s when it was administered by the Bureau of Mines (BOM). With cheap oil supplies and the decline of the coal industry in the 1960s, little funding was allocated to coal cleaning research and development. In 1970, however, the newly created Environmental Protection Agency (EPA) was given funds for this purpose: approximately $1 million in 1970, jumping to about $3.5 million in 1978 and, with a special allocatioh, reaching $9 million in 1979. In 1977, the newly created Department of Energy, which took over the BOM, also began supporting research and development of coal cleaning.[13]

Under the Reagan Administration, all funds for EPA' s coal cleaning budget have been cut beginning in 1982, and DOE's budget is expected to be reduced from $12.6 million in 1980 to $6 million in 1982. The Administration's plan to dismantle DOE make it uncertain whether government funded research will continue at all, although funds may be transferred to the BOM.

Overview of the Coal Cleaning Industry and INFORM's Sample

Coal cleaning operations vary widely in complexity and intensity. They range from crushing and breaking coal pieces to highly intensive sulfur and ash removal. Estimates of the current use of these operations

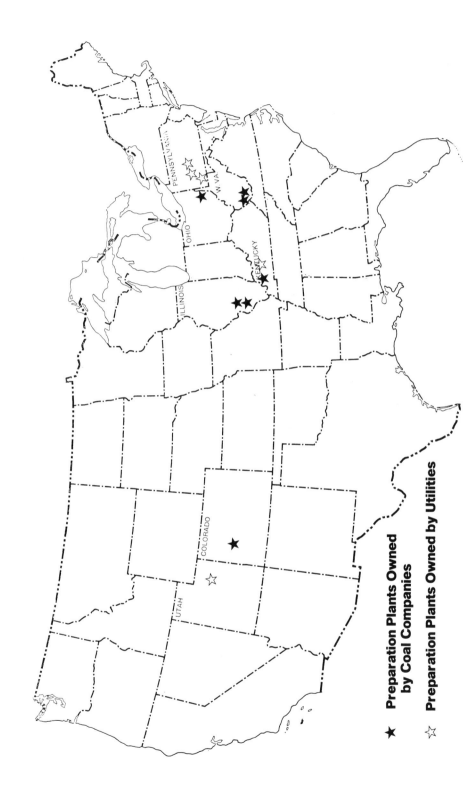

★ **Preparation Plants Owned by Coal Companies**

☆ **Preparation Plants Owned by Utilities**

Table D:
PREPARATION PLANT OWNERS PROFILED

Owner	Plant	Primary Business	Number of Cleaning Plants	Type of Cleaned Coal	Coal Production, 1980 (tons)	Total Production Cleaned(%)
American Electric Power Co.	Helper	Electric utility	9	100% steam	14,058,000	NA
AT Massey Coal Co.	Marrowbone	Coal co. (oil, mineral co. subsidiary)	17	85% steam, 15% metallurgical	12,800,000	NA
Duquesne Light Co.	Warwick	Electric utility	1	100% steam	927,890 (1979)	100
Island Creek Coal Co.	Hamilton	Coal co. (oil co. subsidiary)	28	65% steam, 35% metallurgical	20,000,000	100
Monterey Coal Co.	#2	Coal co. (oil co. subsidiary)	3	100% steam	3,700,000	100
Northwest Coal Co.	Western Slope	Coal co. (energy co. subsidiary)	1	100% steam	600,000	100
Peabody Coal Co.	Randolph	Coal co. (holding co. subsidiary)	18	90% steam, 10% industrial	60,200,000	50
Pennsylvania Electric Co.	Homer City	Electric utility	1	100% steam	(3,258,835(1979)*)	(100)*
Pennsylvania Power & Light Co.	Greenwich	Electric utility	4	100% steam	3,000,000	99
The Pittston Co.	Rum Creek	Coal co.	23	37% steam, 63% metallurgical	25,000,000	80
R & F Coal Co.	Warnock	Coal co. (oil co. subsidiary)	1	100% steam	3,200,000	100
Tennessee Valley Authority	Paradise	Electric utility (government)	4	100% steam	(4,000,000)+	(60)

*Helen and Helvetia "Dedicated Mines"
+TVA does not actually mine its own coal; Peabody Coal Co. mines coal by contract for TVA

across the country do not agree. The National Coal Association, for example, identified 418 operating preparation plants in 1977, while the 1977 Keystone Coal Industry Manual report identified 771 operating plants.

Coal cleaning, particularly the more advanced forms, has been far more prevalent in the east and midwest than in the west, which has only 15 of the preparation plants identified by the National Coal Association. The inherent low sulfur content and high moisture content of western coals has, so far, limited the application of cleaning, although some cleaning of the coals is expected in the future. At the same time, coal production in the west has dramatically increased over the past decade from 35,145,000 tons in 1970 to 135,595,000 tons in 1977.[14] The increase in western coal, and the leveling off of production in the east and midwest, partly explains why the percentage of all coal being cleaned dropped from 53.6 percent in 1970 to 33.8 percent in 1978 (see Table E).[15]*

INFORM looked at four segments of the industry that are most affected by and involved in coal cleaning: 1) companies that clean coal, including mining firms and utilities; 2) architect/engineering/construction (A/E/C) firms that design, engineer and build plants for customers; 3) utilities that burn cleaned coal (both their own and from outside suppliers); and 4) research and development organizations that conduct primary research and development of conventional, advanced physical and chemical processes.

The 12 preparation plants surveyed by INFORM are a representative sample of what coal cleaning can now achieve under a variety of conditions. All of the plants currently clean their own coal for use in electric utility boilers; the company owning the plant, as well as the utility using its cleaned coal are profiled. The preparation plants selected include one owned by an independent coal company, five owned by subsidiaries of oil or energy conglomerates, one owned by a holding company (Peabody), and five owned by electric utilities. Included in our sample are plants owned by Peabody Coal Company, the largest coal producer in the United States and the owner of 20 prepara-

---

*Two other reasons cited account for the fact that the absolute number of tons cleaned dropped over the same years: 1) coke consumption decreased by 20 million tons, and, 2) production from underground mines, which is more often cleaned than surface-mined coal because underground methods are not as selective, dropped from 339 million tons in 1970 to 272 million tons in 1977.[16]

Table E:

MECHANICALLY CLEANED BITUMINOUS COAL AND LIGNITE
(THOUSAND SHORT TONS)

Year	Total Raw Coal Moved to Cleaning Plants	Cleaned by Wet Methods	Cleaned by Pneumatic Methods	Total Cleaned	Total Production	Total Production Mechanically Cleaned(%)	Refuse Resulting in Cleaning Process	Refuse of Raw Coal(%)
1940	115,692	87,290	14,980	102,270	460,771	22.2	13,422	11.6
1950	238,391	183,170	15,529	198,699	516,311	38.5	39,692	16.6
1955	335,458	252,420	20,295	272,715	464,633	58.7	62,743	18.7
1960	337,686	255,030	18,139	273,169	415,512	65.7	65,517	19.4
1965	419,046	306,872	25,384	332,256	512,088	64.9	86,790	20.1
1970	426,606	305,594	17,855	323,452	602,936	53.6	103,159	24.2
1971	361,168	256,892	14,506	271,401	552,192	49.1	89,766	24.9
1972	398,678	281,119	11,710	292,829	595,387	49.2	105,850	26.5
1973	397,646	278,413	10,505	288,918	591,737	48.8	108,728	27.3
1974	363,334	257,592	7,557	265,150	603,406	43.9	98,184	27.0
1975	374,094	260,289	6,704	266,993	648,438	41.2	107,101	28.6
1976	379,236	262,413	6,419	268,830	678,685	39.6	110,403	29.1
1977	358,824	249,447	4,469	253,917	691,344	36.8	104,907	29.2
1978	NA	NA	NA	224,780	665,127	33.8	NA	NA

Source: U.S. Bureau of Mines Yearbook

tion plants, and one owned by American Electric Power, which owns 9 preparation plants, burns more cleaned coal, and produces more electricity than any other U.S. utility.

The sample also demonstrates a range of current technologies, from among the most sophisticated plants in the U.S. today, such as Pittston's Rum Creek Plant, to plants such as Monterey No. 2, owned by an Exxon subsidiary, which cleans its coal moderately. INFORM's sample includes plants of different sizes, from Peabody Coal Company's Randolph plant, which processes the largest tonnage of raw coal of any single plant in the U.S., 2,200 tons per hour, to Northwest Coal Corporation's Western Slope plant, which processes 400 tons per hour.

All ten utilities studied by INFORM use coal cleaned at the preparation plants in our survey. Five of these utilities own their own preparation plants and five buy coal from one or more of the other seven plants.

INFORM analyzed six A/E/C firms that built or retrofitted the preparation plants in our survey. Five of these firms are listed in the 1980 Keystone Coal Industry Manual as the top companies in the coal preparation A/E/C business. A/E/C companies offer a range of services from initial testing of the coal to be cleaned through designing and building the plant to overseeing the start-up of its operations.

INFORM surveyed two types of A/E/C firms. Three companies, Heyl & Patterson, McNally Pittsburg and Roberts & Schaefer, offer services exclusively in coal material handling and plant building and each has developed its own line of coal cleaning equipment. The other three firms, Allen & Garcia, Dravo and Kaiser Engineering are owned by large multinational corporations, all of them highly diversified. None of these firms design or sell preparation plant equipment, although all of them design coal cleaning plants and build other coal related projects such as unit train loading stations, mining shafts, railroads and dams.

Our sample includes three groups of organizations involved in research and development of cleaning processes: those developing conventional coal cleaning processes, those developing advanced physical processes, and those developing chemical processes. INFORM studied the programs of the three major organizations funding coal cleaning research and development: the U.S. Department of Energy, the U.S. Environmental Protection Agency and the Electric Power Research Institute. INFORM studied four companies developing five of the most promising advanced physical processes in depth, and looked at two newer, less certain advanced physical technologies. At least two of these advanced physical methods, the Otisca Process and High Gradient Magnetic Separation (HGMS) are on the verge of commercial use. Finally, INFORM studied nine organizations involved in chemical coal cleaning research and development. All of these organizations, at the time they were chosen, were major developers of the technology and four are presently doing research and development work. All of these companies but one have at some time depended on funding from DOE.

Part II

Findings

Chapter 1

Industry Perspectives

INFORM's study has focused primarily on the technical feasibility of coal cleaning, and on the impact it can have on the cost and efficiency of producing electricity. However, in the course of its research, industry perspectives of the current patterns in the production and use of cleaned coal and on the prospects for coal cleaning in the future have emerged. These perspectives were gathered during interviews with ten utilities, seven coal companies and six architect/engineering/construction (A/E/C) firms.

Utility and Coal Company Views of Current Trends

The utilities interviewed by INFORM do not generally demand cleaned coal. Instead, they set specifications for the amount of sulfur, ash and moisture, and for the Btu content they will accept in the coal they purchase. If suppliers cannot meet utility specifications with their raw coal, they are forced to clean it or look elsewhere for buyers. Every utility interviewed by INFORM stated that, because of the decreasing quality of the coal available, more and more of the coal they purchased is cleaned by their outside suppliers or captive mining operations.

The principal considerations that dictate utility specifications for coal are boiler design and air quality regulations. In order to insure that electricity is produced as efficiently as possible, utilities seek a coal that will match the design of their boilers, or they design boilers to suit the best available coal at their site. After 1970, utilities have also had to consider what quality of coal to burn to help them meet emission standards under the Clean Air Act.

Beyond meeting boiler design and air quality requirements, cost is the determining factor used by most utilities when obtaining coal of the quality they need, according to the utilities in INFORM's sample. To find the least expensive coal, a utility may have to weigh the cost of buying and cleaning a local high sulfur coal against the cost of transporting a more distant coal, perhaps containing mining wastes, that is low in sulfur and ash.

According to the coal companies and utilities surveyed by INFORM, two factors have encouraged more intensive coal cleaning: the Clean Air Act and the rising price of coal.

Moderate to extensive cleaning has enabled many boilers burning medium-sulfur coals to meet air quality standards set before 1977. It has also, in combination with scrubbers, helped other boilers comply with the stringent 1977 Revised New Source Performance Standards.

In the past, it was frequently not economical to clean the very smallest sized coal particles, and they were often thrown away as waste or added raw to coal that had been cleaned. However, the average price, excluding transportation, for a ton of coal rose from $9.20 in 1970 to $20.55 in 1977. Many companies are now building plants with equipment to clean and recover these finer-sized particles.

Island Creek, one of the coal companies interviewed by INFORM, has retrofitted its older cleaning plants with new equipment (hydrocyclones) in order to meet customer demand for cleaner coal. Two other coal companies, Pittston and Peabody, reported that the plants they build in the future will be designed to clean coal more extensively, in large part because of more stringent specifications set by utilities.

While the Clean Air Act has stimulated more intensive coal cleaning, the utilities and coal companies in INFORM's survey resist the idea of being required to burn cleaned coal. Utilities want to have choices about how to meet air quality standards; for example, whether their boilers built between 1971 and 1978 meet the federal New Source Performance Standards of 1.2 lb $SO_2/10^6$ Btu by using low-sulfur coal (either raw or cleaned) or by using scrubbers and high-sulfur coal.

A statement by one utility spokesman reflects the opinion of many:

> All coal does not need to be cleaned. Carolina Power and Light purchases some high quality coal that does not require washing. Coal cleaning should remain an option offering a utility flexibility in meeting air quality standards, rather than another requirement for an industry already overburdened with regulations.

Peabody Coal Company, which cleans approximately 50 percent of its annual production, also serves this caution: "One cannot make generalizations about the advantages of coal cleaning across the board. Cases are site specific. Coal seams are different. Buyer requirements are different."

An increasing number of utilities own coal reserves, and according to those interviewed by INFORM, more and more are expressing an interest in building their own preparation plants. Production of coal from utility-owned mines has risen over the past fifteen years, from 4.5 percent in 1965 to 11 percent in 1979. The Federal Regulatory Commission predicts that by 1985, utility-owned mines will account for 21 percent of the coal market. The 13.5 billion tons of recoverable reserves now owned by 82 of the nation's 239 investor-owned utilities make them the fifth largest coal owning group.[1]

According to half of the utilities interviewed by INFORM, utility ownership and management of mines will continue to grow for two principal reasons. First, with the decline in coal quality and the expected increase in demand for coal, control over the production and preparation of coal represents a good investment for utilities. It allows them to be sure that they get the quality of coal they need over a long period of time, at a price they hope they can more readily predict. Second, utilities often have access to more capital to invest in preparation plants than do independent coal companies.

Five of the ten utilities INFORM studied own their own preparation plants. One utility, AEP, has a preparation plant at every one of its seven coal mining subsidiaries, and two plants at one of them. Another, Associated Electric, bought a Missouri mine from Peabody Coal Company that had coal with such a high sulfur and ash content that the utility has invested in a cleaning plant which is to be completed in 1982. A third, PENELEC, has recently completed its Homer City preparation plant, so that coal from its reserves can meet a 1.2 lb $SO_2/10^6$ Btu standard.* TVA, which also owns reserves, has just completed its Paradise preparation plant to meet a 5.2 lb $SO_2/10^6$ Btu standard with coal of relatively poor quality that it has under existing supply contracts.

Most coal company incentives for cleaning coal are based on improving their competitive edge in a cyclical coal market. According to the coal companies in INFORM's sample, cleaning provides one or a combination of the following advantages: a high quality coal that can frequently be obtained by washing at less expense than is provided by "selective mining;" a coal that would otherwise be too dirty to sell; greater flexibility in looking for buyers; and the ability to offer "compliance coal," designed to meet air quality standards.

Northwest Coal Company and Associated Electric Cooperative gave up "expensive and less efficient selective mining techniques" and built their preparation plants to reduce the ash and mining wastes in their coal. Associated Electric Cooperative reports that selective mining slowed production by about 30 percent; it plans to stop this practice altogether when its preparation plant is completed. Even with selective mining, Northwest Coal discovered that most utilities would not buy its low sulfur, high Btu western coal because of the coal's high ash content. By removing ash, cleaning will satisfy utility buyers. By increasing the coal's already high Btu value from 12,000 to 12,700 per pound, cleaning will also allow the company to sell its coal for industrial stoker-fired boilers, which demand a coal of very high quality (about 13,000 Btu's per pound, 0.65 percent sulfur, 5 percent ash and 5 percent moisture content).

Pittston sells a compliance coal guaranteed to meet a specific sulfur dioxide ceiling when burned. This coal can be used in utility boilers built before September 1978 which are subject to standards less stringent than the current federal Revised New Source Performance Standards (RNSPS) requiring 90 percent sulfur removal.

One disincentive to the building of preparation plants is their capital costs, which in INFORM's sample ranged from $2.2 million to $97.0 million, primarily depending upon the capacity and complexity of the plant; the majority cost between $10 and $25 million (see Findings: Coal Cleaning Plants and What They Can Do). The high 1981 in-

---

*PENELEC has not yet met this standard with coal from its Homer City plant, although they project that they will achieve compliance by March, 1987.

terest rates of 20 percent or more on borrowed money have compounded this expense. According to the A/E/C firms and coal companies interviewed by INFORM, only utilities, subsidiaries of oil or energy companies, and large coal companies can meet such expenses. Smaller, independent mining operations may not have the financial resources to build preparation plants.

Future Prospects

Three of the six A/E/C firms, and all of the seven coal companies and ten utilities interviewed by INFORM predict more cleaning of coal. The other three A/E/C firms find coal cleaning's future clouded by the changes and unpredictability of federal regulatory and funding policies.

In addition to the declining quality of the nation's coal reserves and the air quality regulations governing power plant emissions, the companies in INFORM's sample cited two additional reasons for an increased cleaning of coal: the growth anticipated in the demand for steam coal for export, and the development of new energy technologies that use cleaned coal.

Three of the coal companies studied by INFORM are deeply involved in the coal export market. In 1980, A.T. Massey, Island Creek and Pittston (the largest U.S. coal exporter) shipped 6 million, 5.4 million and 11.2 million tons of coal respectively to customers in Europe, Asia and South America. Two of them (A.T. Massey would not grant INFORM an interview) agree with the World Coal Study and other projections that increasing amounts of American steam coal will be exported in the next ten to twenty years. They also predict that foreign buyers will demand cleaned coal because they will not pay to ship large amounts of impurities that could be removed.

Two coal companies also believe that new energy technologies will encourage the use of cleaned coal. Pittston, for example, predicts a market for its cleaned coal with utility and industrial users of coal-oil mixtures (COM), which are now being developed by a number of organizations, including Florida Power and Light Company and the New England Electric System at its Salem plant. The future of synthetic fuels is less certain, but if this industry does become commercially viable in the next ten years, R & F Coal Company believes that it will provide a market for cleaned coal. However, others in the industry are less certain of the role of coal cleaning in synfuels production.[2]

All the companies interviewed by INFORM made the point that federal policies will continue to have a strong impact on coal use, and consequently on the amount of coal that is cleaned. Two coal companies, Pittston and Island Creek, complained that the federal government's promised promotion of increased coal use has not been realized. For example, few of the orders issued under the 1978 Power Plant and Industrial Fuel Use Act (FUA) (now repealed), for power plants to convert from oil to coal were ever carried out. Unless the federal government more actively promotes coal use or unless oil prices rise dramatically, these companies believe that the market for coal will grow more slowly than predicted three or four years ago.

Companies also asserted that federal regulatory policies will affect the coal and coal cleaning markets. Some want the Clean Air Act amended to reduce emissions restrictions. Pittston, for example, believes that air quality regulations have limited the growth of the coal market. All the utilities in INFORM's sample believe there should be more flexibility in meeting the standards. It is also clear that predictability and consistency are important to all users as a basis for any viable planning.

Chapter 2

Coal Cleaning Plants and What They Can Do

The size and complexity of the 12 coal preparation plants sampled by INFORM vary widely, as do the kinds of coal used, methods of preparation and the degree to which the coal is cleaned. A range of factors influences the operations and costs of each plant, and no two pieces of coal or no two plants are alike.

Listed below are the key findings from INFORM's sample of 12 coal preparation plants, discussed further in this chapter:

The *average sulfur removal* was 25 percent and the average ash removal was 51 percent. The key factors determining how much sulfur and ash can be removed include the nature of the raw coal feed, the intensity and method of preparation, and the amount of money a given plant owner can afford to spend to achieve optimum cleaning results.

Cleaning increased the Btu or heating value of the raw coal by 5 to 44 percent, depending upon the nature of the raw coal and the intensity of the process. Although the removal of ash and mining wastes affected Btu improvements, in most cases coal cleaning increased the moisture content of the coal unless thermal drying was employed.*

---

*Thermally drying coal to reduce moisture has alleviated this problem in five plants profiled by INFORM. However, thermal drying is energy intensive and therefore costly.

The preparation plants ran an *average of two eight-hour production shifts* and one eight-hour maintenance shift daily. The average "downtime" ranged from 2 to 10 percent of the time the owners wanted the plants to operate. Downtime was primarily related to cleaning equipment and conveyor breakdown, instrument and control failure, and worker maintenance difficulties. The most downtime was reported when plants were in the "start-up" phase of operations.

The maximum amount of *power needed* to run preparation plants ranged from 0.4 to about 20 megawatts of electrical power. Power requirements depended on plant size, complexity and whether or not thermal drying was employed. The plant's operators did not provide information on the costs of power, especially in terms of the costs per ton of processed coal.

The amount of *water used* varied considerably, and depended not only on the plant's size and complexity, but also on plant design factors. Most of the plant owners considered the cost of water to be "minimal," and described few problems in obtaining it for coal cleaning. All the plants recycle their water, although some is lost through evaporation and absorption. Make-up water needs ranged from 87 to 2000 gallons per minute.

Daily *waste production* ranged from 1,316 to 8,384 tons. Most plants dispose of their wastes in landfills designed to accept refuse for the life of the plant or associated mine (up to 40 years). Other waste disposal options were slurry ponds and dams constructed from preparation plant wastes.

The *construction of preparation plants* required between ten months (Western Slope plant) and three and one-half years (Rum Creek plant), depending on the plant's size, complexity and the time needed to comply with environmental regulations. This amount of time, according to the coal companies and the A/E/C firms profiled by INFORM, should not be a factor limiting the expanded demand for cleaned coal.

The *capital costs* of preparation plants ranged from $2.2 million (1979 dollars) at the small Western Slope plant to $97.2 million (1981 dollars) at the Homer City plant, which has required extensive retrofitting.

Table A: INFORM'S SAMPLE OF PREPARATION PLANTS

Owner/ Plant	Location	Size (tph)*	Level	Major Cleaning Equipment Used	Coal Seam(s) Cleaned
AEP Helper	Helper, UT	1,100	4	Heavy-media washers; Heavy-media cyclones, Water-only cyclones; froth flotation units	Subseam No. 3
AT Massey Marrowbone	Naugatuck, WV	NA	4	Heavy-media cyclones; froth-flotation units	Coalburg
Duquesne Warwick	Greensboro, PA	900	3	Heavy-media washers, Water-only cyclones	Sewickley
Island Creek Hamilton	Morganfield, KY	800	4	Heavy-media cyclones	Western No. 9
Monterey No. 2	Albert, IL	1,350	3	Jigs	Illinois No. 6
Northwest Coal Western Slope	Somerset, CO	400	3	Jigs	Hawksnest
Peabody Randolph	Marissa, IL	1,676	4	Jigs, water-only cyclones	Illinois No. 6
PENELEC Homer City	Homer City, PA	960	5	Heavy-media cyclones; Water-only cyclones, Deister tables	Upper Freeport-E Lower Freeport-D
PP&L/PMC Greenwich	Greenwich, PA	900	4	Jigs	Lower Freeport
Pittston Rum Creek	Lyburn, WV	500	4	Heavy-media washers; Heavy-media cyclones; Water-only cyclones; froth-flotation units	Coalburg, Stockton
R&F Warnock	Warnock, OH	800	4	Heavy-media washers, Heavy-media cyclones; Water-only cyclones	Pittsburgh No. 8 Meigs Creek No. 9 Waynesburg No. 11
TVA Paradise	Drakesboro, KY	1,607	4	Heavy-media washers; Water-only cyclones; Heavy-media cyclones; Froth flotation units	Muhlenburg Co. Nos. 9, 11, 12

*tons per hour; cleaned coal output

The Effects of Coal Cleaning on Sulfur, Ash, Moisture and Btu Content

Physical coal cleaning can remove varying portions of a coal's sulfur and ash. Eight of the twelve preparation plant owners sampled provided INFORM with the sulfur and ash content of their raw and cleaned coal. The following table shows that these plants remove about 25 percent of the sulfur from raw coal, and about 51 percent of the ash. An exceptional case in our sample, PENELEC's Homer City preparation plant, has been built to clean its coal more extensively than other U.S. coal cleaning plants. (See Homer City profile.) According to PENELEC, the plant will remove up to 75 percent of the sulfur and 86 percent of the ash once the plant's coal cleaning equipment is retrofitted in February 1982.

The amount of sulfur and ash removed by a particular plant depends on several factors. One of the most important is the nature of the raw coal feed. For some raw coals, the amount of sulfur and ash, the type of sulfur and the size of the sulfur particles make sulfur and ash difficult to remove with the physical washing technologies, which do not alter the molecular structure of the coal—as do the chemical coal cleaning processes. In coals such as western bituminous and sub-bituminous, a high percentage, up to 66 percent, of the sulfur is organic and only 34 percent is inorganic.* Organic sulfur cannot be removed by physical cleaning methods. For example, Northwest Coal's Western Slope plant removes ash, but does not remove sulfur from its raw low-sulfur western coal.

Even inorganic sulfur particles cannot always be economically removed if they are very fine-size and widely disseminated in the raw coal. The intrinsic ash, comprising only a small portion (less than 2 percent) of the ash, (see Appendix: Description of Coal) also cannot be removed by cleaning, since it is part of the structure of the coal.

The "level" of coal preparation (see Methods of Coal Preparation section for a description of preparation plant levels) is another important factor that determines how much sulfur and ash will be removed. The plants profiled by INFORM ranged from facilities producing "moderately" cleaned coal (level 3) to plants that produce two grades

---

*On the other hand, Northern Appalachian coals contain an average of about 67 percent inorganic and 33 percent organic sulfur, according to the U.S. Bureau of Mines.

of coal, one moderately cleaned and one intensively cleaned to level 5. As a rule, more sulfur and ash is removed in the higher-level plants where raw coal is crushed to smaller sizes and cleaned in more sophisticated equipment (i.e., in heavy-media cyclones and froth-flotation units).

Other factors, determined by customers' particular fuel needs, also dictate the amount of sulfur and ash removed. For example, some plants are designed to clean coal so that a high percentage (90 to 95 percent) of the Btu value of the raw coal is returned. These plants attempt to recover burnable coal particles by keeping them out of the waste stream. However, by recovering these burnable particles, the plant may also retain some of the sulfur and ash in the coal that could otherwise have been removed. In determining how much to clean its coal, plant operators balance their customers' needs for sulfur and ash removal against their own need to recover Btus.* In addition, plant operators must determine if the added expenses of coal cleaning can be recouped from their customers through higher coal prices; in other words, are the customers willing to pay the higher price for higher quality coal?

In general, more Btus are lost in the wastes of cleaned coal from which a high percentage of ash has been removed, than are lost in the wastes of coal subjected to less cleaning. Duquesne's Warwick plant, for example, recovers 96 percent of the Btu value of its raw coal. This high percentage of Btu recovery is a priority for the utility, which seeks to minimize the loss of coal in its coal cleaning plant wastes. However, the plant removes only 50 percent of the ash from the raw coal, a removal rate acceptable to the utility. On the other hand, the Monterey No. 2 plant removes 68 percent of the ash from its raw coal, but the cleaned coal has only about 90 percent of the Btu value that the same coal had before cleaning. (Table B lists the Btu content of the raw and cleaned coal for seven of the twelve preparation plants profiled by INFORM.)

The water used in most coal cleaning processes is adsorbed as surface moisture by the cleaned coal, decreasing its net Btu value, and partially offsetting the increase in heating value achieved by waste removal. Most buyers of cleaned coal set specifications so that the fuel does not exceed a maximum moisture content. The percent increase in

---

*The heating value of the cleaned coal is 5 to 44 percent greater than that of the raw coal for an equivalent weight.

Table B: SULFUR, ASH, AND BTU RECOVERY (MOISTURE-FREE BASIS)

Owner/ Plant	SULFUR CONTENT (%)			ASH CONTENT (%)			Btu Recovery(%)
	Raw Coal	Clean Coal	Reduction	Raw Coal	Clean Coal	Reduction	
AEP Helper	NA	NA	NA	NA	NA	NA	NA
AT Massey Marrowbone	NA	NA	NA	NA	NA	NA	NA
Duquesne Warwick	2.30	1.90	17.4	32.00	15.90	50.3	96
Island Creek Hamilton	3.85	2.80	27.3	21.84	8.90	59.3	96.5
Monterey No. 2	3.90	3.30	15.4	25.00	8.00	68.0	$\pm$90
Northwest Coal Western Slope	0.60	0.60	0.0	20.00	8.00	60.0	$\pm$90
Peabody Randolph	NA	NA	NA	NA	NA	NA	NA
PENELEC Homer City	3.40	2.35* 1.81+	31.0* 46.9+	20.00	17.75* 8.00+	11.3* 60.00+	95.4
PP&L/PMC Greenwich	1.50 to 2.50	1.40 to 1.70	+20.0 (average)	25.00 to 33.00	15.00 (maximum)	40.0 to 55.0	96.1
Pittston Rum Creek	NA	NA	0.15 (increase)	NA	NA	24.0	NA
R&F Warnock	2.00 to 6.00	1.25 to 5.25	37.5 to 12.5§	12.00 to 20.00	8.00 to 9.00	33.3 to 55.0**	§
TVA Paradise	4.90	3.10	36.7	20.10	9.00	55.2	93

*Medium-Cleaned Coal - see PENELEC Profile
+Extensively-Cleaned Coal
**Percent reduction calculated by assuming reduction from top to bottom range of sulfur and ash specifications
§Because R&F's raw coal comes from over six seams and pits, its Btu content is too variable to calculate the percentage of Btus recovered in the cleaned coal

Table B: MOISTURE AND BTU/LB (MOISTURE FREE BASIS)

Owner/ Plant	MOISTURE CONTENT (%)			BTU/LB		
	Raw Coal	Cleaned Coal	Increase/Reduction	Raw Coal	Cleaned Coal	Increase (%)
AEP Helper	NA	NA	NA	NA	NA	NA
AT Massey Marrowbone	NA	NA	NA	NA	NA	NA
Duquesne Warwick	3.0 to 4.0	6.3	+57.5 to 110.0	+8,000	+11,500	+43.8
Island Creek Hamilton	9.2	9.4§	+2.2	NA	NA	NA
Monterey No. 2	13.5	15.5	+14.8	8,400	10,800	28.6
Northwest Coal Western Slope	5.0	7.5	+50.0	12,000	12,400	3.3
Peabody Randolph	NA	NA	NA	NA	NA	NA
PENELEC Homer City	6.0	4.0§	-33.3	+11,800	12,549* 14,000+	6.4* 18.6+
PP&L/PMC Greenwich	4.0 to 5.0	7.0§	+40.0 to 75.0	+10,300	+12,500	+21.4
Pittston Rum Creek	NA	NA§	0	NA	NA	NA*
R&F Warnock	6.0	8.0§	+33.3	10,500 to 12,000	12,000 to 14,000**	14.3 to 16.7
TVA Paradise	10.7	12.2	+14.0	11,407	13,206	15.8

*Medium-Cleaned Coal - See profile of Homer City
+Extensively-Cleaned Coal Plant
**% Btu increase calculated by assuming increase from bottom to top of Btu range
§Thermal dryers used

moisture content (after shipping) of the cleaned coal over the raw coal in seven of the eight preparation plants that reported moisture specifications ranged from 2.2 percent to 11.0 percent.

One reason that western sub-bituminous and lignite coals are poor candidates for cleaning is because of their inherently low sulfur and ash content, and high inherent moisture content.* The wet coal cleaning technologies that add moisture to the coal can offset the benefits gained by removing the solid impurities since the main contaminant of these coals is water.

Five plants in INFORM's study use thermal dryers to decrease the moisture content of their cleaned coal. One of them, the Homer City facility, actually reduces the moisture content of its cleaned coal by 33 percent relative to the raw coal. Thermal dryers are not used by many preparation plants because they are energy intensive and thus costly— partly as a result of the necessity of installing scrubbers on them to control particulates.

Operation and Maintenance

The companies in INFORM's sample report that an average of one-third of a preparation plant's working time is devoted to maintenance. Because preparation plants use machinery that contains many moving parts subject to constant wear and tear, plant owners state that they must schedule daily monitoring and maintenance of equipment if production efficiency is to be kept high. (Table C lists maintenance schedules reported by ten of the twelve preparation plants in INFORM's sample.)

These plant owners reported production shifts from 14 to 24 hours a day, and maintenance shifts from 20 to 88 hours per week. On the average, there was one 8-hour maintenance shift for every 16 hours (or two 8-hour production shifts) of coal cleaning production.

Many of the plants also reported that additional maintenance activities were performed at times other than the 8-hour maintenance shifts, either on weekends or during vacations. In addition, increased maintenance is usually necessary during the "start-up" of new plants.

---

*In addition, most western sub-bituminous coals contain high percentages of sodium. This substance, which increases the problem of slagging and fouling in boilers, is often concentrated when coal is cleaned.

The most maintenance time reported was at PENELEC's Homer City preparation plant, which required an exceptionally high start-up maintenance period of 88 hours per week (44 hours per cleaning circuit) between January and May of 1980. PENELEC attributed this additional maintenance to difficulties encountered with design, equipment and personnel when starting up this very complex plant (see Homer City profile).

The average downtime of the four preparation plants that responded to questions about downtime percentages ranged from 2 to 10 percent of the operation time, i.e., the time the owners wanted the facility to operate. The downtime was attributed mainly to coal cleaning equipment and conveyor belt breakdowns, plant instrument and control equipment failures, and the inability of plant workers to identify and correct equipment problems quickly. Conveyor systems and thermal dryer control equipment also tended to require more maintenance than other pieces of plant equipment.

Preparation plant owners have developed various methods of coping with problems of downtime and maintenance. At the R&F Coal Company's Warnock plant, coal tended to clog conveyors and some of the cleaning equipment during wet, cold weather. To correct this, R&F doubled the capacity of the crushers used before the raw coal enters the cleaning circuits, so that the coal which entered the plant in large slabs from three stockpiles was crushed first.

Other plants, such as the Peabody Coal Company/Randolph plant, reduce downtime by using two independent cleaning circuits. This enables one cleaning circuit to operate while the other is being maintained or repaired. Another practice that reduces the impact of plant downtime on coal deliveries is used by Duquesne Light Company's Warwick plant, which stockpiles about 11,000 tons of cleaned coal at the plant site. This coal is shipped to utility-owned power plants when the preparation plant is shut down.

Finally, in-house training programs for plant operators and mechanics have been used at the PENELEC and PP&L/PMC preparation plants to reduce unacceptable periods of downtime caused by inexperienced workers. PP&L/PMC reports that their training program has helped to cut downtime and equipment outages at the company's Greenwich plant, and that "the program has already paid for itself."

Table C: OPERATION AND MAINTENANCE OF PREPARATION PLANTS

Owner/ Plant	Normal Plant Operations	Scheduled Maintenance	Average Downtime
AEP Helper	NA	NA	NA
AT Massey Marrowbone	NA	NA	NA
Duquesne Warwick	14 hours per day; 5 days per week; 50 weeks per year*	7¼ hours per day; +40 hours per week	+2% of the time that Duquesne wanted to operate the plant
Island Creek Hamilton	13⅓ hours per day; 4.7 days per week; 235 days per year	40 hours per week	5% of total work time
Monterey No. 2	16 hours per day; 235 days per year	8 hours per day	NA
Northwest Coal Western Slope	14 hours per day; 5 days per week	+80 hours per week; ⅚320 hours per month	10% of total work time
Peabody Randolph	16 hours per day	8 hours per day	NA
PENELEC Homer City	24 hours per day; 5 days per week; 50 weeks per year+	88 hours per week	"High," due to plant start-up and design difficulties
PP&L/PMC Greenwich	14 hours per day; 5 days per week; 50 weeks per year	8 hours per day; 40 hours per week	NA
Pittston Rum Creek	16 hours per day; 5 days per week; 50 weeks per year	60 "man hours" per week	NA
R&F Warnock	20 hours per day; 5 days per week**	4 hours per day; 20 hours per week	NA
TVA Paradise	16 hours per day; 5 days per week; 52 weeks per year	8 hours per day	NA

*less holidays
+figure includes maintenance time
**or until weekly quota of 50,000 tons is produced

Use of Power, Water, Magnetite and Reagents

Nine of the twelve preparation plants profiled by INFORM furnished information concerning at least a portion of their resource requirements (power, water, magnetite and cleaning reagents).* However, the information concerning the consumption of energy and water in relation to the amount of coal cleaned was limited. Furthermore, data was often provided by different companies in different units of measurement, making it difficult to compare. (All of this data is presented in Table D.)

Figures provided by seven plant owners, referring to the maximum amount of electrical power needed to run the plant at full capacity, were usually expressed in megawatts (Mw) or horsepower (Hp). These ratings ranged from 0.4 Mw at the small, relatively simple Western Slope plant, to over 20 Mw at the larger, more complex Rum Creek and Homer City plants. A more useful figure for analyzing the energy efficiency of different coal cleaning systems would be their electricity consumption, or power consumption, per ton of cleaned coal. Such a figure was not provided by any of the plants in INFORM's sample; some plant owners said that they did not know this figure and others claimed that such information was proprietary.

The preparation plant owners in INFORM's sample provided figures describing the amount of water circulating in their plants in different units of measurement. For example, water consumption or use was described in "gallons used per minute," "gallons used per day," "gallons per ton of coal" and "gallons of make-up water."

Because regulations established under the Federal Water Pollution Control Act (also known as the Clean Water Act) limit water pollutants discharged from coal preparation plants, all of the plants profiled by INFORM employ a "closed water loop" plant design. (See Appendix for a full discussion of preparation plant waste disposal practices.) All of the water used to process cleaned coal and waste materials is clarified and recycled for reuse in the plant (for a description of water clarification systems, see Methods of Coal Preparation section). However, a

---

*The three cleaning reagents most commonly used in preparation plants are: (1) flocculants used in water clarification equipment, (2) pH control reagents such as lime, and (3) reagents used in froth flotation equipment, e.g., fuel oil.

certain percentage of the water is "lost" through the evaporation, and adsorption by the coal and refuse. Also, waste water may be removed from the cleaning circuits in order to be clarified, further reducing the amount of water circulating through the plant. "Make-up" water must therefore be added to the plant cleaning circuits to replace this water loss.

Make-up water needs for the six plants in INFORM's sample that provided such information ranged from 87 to 2000 gallons per minute. The large and more complex plants usually needed larger amounts of water than the smaller, less complex plants. However, no rough ratio exists between water needs and plant size. For example, the 1,607 ton per hour TVA/Paradise preparation plant only requires 195 gallons of make-up water per minute, which is a small quantity relative to the plant's size. Where water is in short supply and where coal cleaning plants compete for this resource with other water users, the amount of make-up water required per ton of cleaned coal would be especially important information. In addition, plants that clean fine-size coal particles generally require greater amounts of water than plants that clean larger sized coal particles.

Three of the plants reported using between 250 and 420 tons of magnetite per month in their heavy-media vessels or cyclones. Only two plants reported magnetite consumption in terms of pounds required per ton of raw and/or cleaned coal. Although there is little basis for comparison in INFORM's sample, the PENELEC/Homer City plant did report that its usage (two to seven pounds of magnetite per ton of cleaned coal) was "unusually" large relative to other plants. PENELEC attributed its excessive use of magnetite to problems associated with the equipment used to reclaim the magnetite from cleaned coal and plant wastes. Magnetite is not a rare mineral but excessive use can add to the cost of coal cleaning. Island Creek Coal Company was the only plant owner that provided INFORM with their magnetite costs. They reported that the cost of magnetite at the company's 800 ton per hour Hamilton Preparation Plant* (which uses about 420 tons of magnetite per month) was $0.05 per pound or about $42,000 per month (1979).

Only four preparation plants of the 12 in INFORM's sample use froth-flotation equipment to clean fine-size coal particles. Of the three

---

*Hamilton is a representative preparation plant that uses magnetite in its heavy-media cyclones.

Table D: PLANT RESOURCE CHARACTERISTICS

Owner/ Plant	Electrical Power Rating (Mw)	(hp)	Water	Magnetite/Reagent Use
AEP Helper	NA	NA	NA	NA
AT Massey Marrowbone	NA	NA	NA	NA
Duquesne Warwick	NA	NA	125 gallons/minute added as make-up water	1.2 lbs magnetite per ton of raw coal for heavy-media washers
Island Creek Hamilton	6.7	9,000	250-300 gallons/minute added as make-up water	420 tons magnetite per month for dense-media cyclones
Monterey No. 2	5.6	7,500	2,000 gallons/minute added as make-up water	None used
Northwest Coal Western Slope	0.4	550	87 gallons/minute added as make-up water; 13 gallons of recycled water per ton of coal	None used
Peabody Randolph	NA	NA	8 gallons of recycled water per ton of coal	NA
PENELEC Homer City	12 (20 at full capacity)	16,100	16,000 gallons per minute circulate in plant; make-up water added as needed	2-3 lbs per ton of moderately cleaned coal; 5-7 lbs per ton of extensively cleaned coal
PP&L/PMC Greenwich	5 (6 when loading train)	6,700	10 gallons of recycled water; 7 to 14 million gallons circulate per day	None used
Pittston Rum Creek	20.6	27,600	650 gallons of recycled water per ton of coal	250 tons magnetite per month for heavy-media vessels & cyclone; 0.4 lbs reagent and fuel oil per ton of raw coal for froth flotation units
R&F Warnock	NA	NA	600 gallons per minute added as make-up water	NA
TVA Paradise	10* (installed load)	13,400	195 gallons per minute added as make-up water	Expect to use 320 tons of magnetite per month in heavy media equipment; 27 lbs per ton of raw coal for heavy media equipment

*2 to 3 Mw is the plant's normal load

plants, only the Pittston/Rum Creek facility provided data on reagent and oil use in the froth-flotation equipment (as can be seen in Table D).

Waste Disposal

Daily waste production as reported by 10 of the 12 plant owners profiled ranged from 1,316 tons at Northwest Coal's Western Slope plant to 8,384 tons at Peabody's Randolph plant. Five major factors determine the amount of waste a plant produces: 1) the amount of coal processed, 2) the percentages of ash and mining wastes in the raw coal, 3) the percentage of waste removal achieved, 4) the number of hours per day the plant operates, and 5) the type of cleaning equipment used.

Eight of the ten plants profiled by INFORM that responded to questions about waste disposal reported that their wastes are disposed of in landfills. These eight plants bury at least part of their coarse and fine wastes, which are typically trucked or conveyed to a nearby site where they are spread and compacted. Waste compaction encourages the runoff of precipitation, which can be collected and treated to reduce its acidity, the concentration of suspended solids and trace element pollution (see Appendix: Coal Preparation Waste Disposal).

Landfill sites are usually designed to accept wastes for the projected life (up to 40 years) of the preparation plant or its associated mine. Two plant owners provided information on the size and expected capacity of their waste disposal sites. At the present time, PP&L/PMC's Greenwich plant disposes of its 3,600 tons of daily wastes in an 80 acre landfill. The company plans to develop a 100 acre site in the mid-1980s that is projected to receive wastes for the remaining 30 years of operations at its mine and preparation plant. TVA purchased 2,600 acres adjacent to its Paradise plant in which it will dispose of the 4,528 tons of wastes produced daily for the life of the plant (20 to 30 years).

The use of "closed water loop" designs (discussed above) has reduced the volume of fine-waste slurries at many plants. In newer plants, fine wastes are removed from waste streams to be concentrated in static thickeners and filter presses, and the clarified waste water is reused in the cleaning circuits.* The fine wastes are either buried or

---

*PENELEC's Homer City plant uses ponds to collect fine wastes that are present in water runoff from the plant site. These wastes are later disposed of in a landfill.

disposed of in slurry ponds. For example, Island Creek's Hamilton 1 plant currently uses slurry ponds for fine waste disposal. The fine waste is allowed to settle to the bottom of the pond (or impoundment) and the relatively clear waters from the top of the ponds are pumped back to the preparation plant for reuse. The fine wastes are either allowed to accumulate in the ponds until the structures are "full," or removed and buried along with the coarse wastes.

Table E: WASTES PRODUCED AT PREPARATION PLANTS

Owner/ Plant	Amount of Wastes Produced	Wastes Recovered From Raw Coal Feed (%)
AEP Helper	NA	NA
AT Massey Marrowbone	NA	NA
Duquesne Warwick	1,885 tons when 3,500 tons of cleaned coal are produced	+35
Island Creek Hamilton	3,500 tons per day; 50% coarse, 50% fine	30
Monterey No. 2	+6,172 tons per day	30 to 35
Northwest Coal Western Slope	1,316 tons per day; 85% coarse, 15% fine	20 to 25
Peabody Randolph	8,384 tons per day	24
PENELEC Homer City	5,000 tons per day at full capacity	+20
PP&L/PMC Greenwich	3,600 tons per day; 50% coarse, 50% fine	+28
Pittston Rum Creek	5,200 tons per day	40
R&F Warnock	+2,890 tons per day; +14,448 tons per week	23
TVA Paradise	4,528 tons per day	15*

*Wastes may constitute up to 20 percent of the raw coal feed

There is only one other waste disposal option being exercised by preparation plant owners in INFORM's sample. The R&F Coal Company's Warnock plant is currently disposing of its coarse plant wastes by using them to build a dam, which serves to clarify and hold the make-up water supply.

Estimates of waste disposal costs were only provided by Duquesne Light Company/Warwick and Northwest Coal/Western Slope. In 1979, waste disposal costs for the Warwick plant were $1.47 per ton of cleaned coal. This was within the range of costs reported by published studies of preparation plant wastes (see Appendix: Coal Preparation Plant Waste Disposal, for a discussion of these studies). Northwest Coal, which now transports its refuse 26 miles to a county landfill, reported high waste disposal costs of $3 per ton of refuse.

Few of the preparation plant operators profiled commented on the environmental effects of plant operations not associated with waste disposal. PENELEC's Homer City plant reported that "fugitive" dust was emitted from conveyors that transported fine coal particles. The problem has been solved by building metal covers over conveyor belts and using dust control sprays.

Particulate emissions have been controlled by installing scrubbers in the thermal dryers used to decrease the moisture content of fine coal. This has been done at three of the plants profiled (PENELEC/Homer City, PP&L-PMC/Greenwich, and Pittston/Rum Creek). Utility power plants have used these same technologies.

Time Needed for the Building of Preparation Plants

Plans to significantly increase the use of cleaned coal must take into account the time required to build a preparation plant and the variables that may change it. Among the nine plants that provided information on this subject to INFORM, construction time from design through building to start-up ranged from ten months (for Northwest Coal's Western Slope preparation plant) to three and one-half years (for Pittston's Rum Creek plant) (see Table F). Close examination of these two plants illustrates some of the key factors that influence the length of time needed for plant construction.

Both the Western Slope and Rum Creek plants were designed to process similar amounts of raw coal, 400 and 500 tons per hour respectively. Western Slope's raw coal comes from a single mine. The plant does not need to reduce the already low sulfur (0.6 percent) content of its coal, and is only required to decrease the ash content of its raw coal while increasing its Btu content. For these purposes, a relatively simple piece of equipment, a Baum jig, is used.

In contrast, the Rum Creek plant, built for the Pittston Company by the McNally Pittsburg Company, was designed for much greater flexibility and more intensive cleaning than Western Slope, and uses more sophisticated cleaning equipment. The plant was designed to reduce the sulfur and ash content of coal mined in five different locations, where the quality of the raw fuel may vary significantly. The sulfur dioxide emissions generated when cleaned coal from Rum Creek is burned are guaranteed to meet a $1.2/10^6$ Btu emission limit. More sophisticated plant equipment (i.e., heavy-media and froth-flotation units), thermal dryers, coal quality testing instruments, and automated plant circuitry were necessary to accomplish this higher level of coal cleaning.

In addition, the Rum Creek plant was designed to comply with present and future environmental and safety regulations. Equipment such as dry dust collectors, thermal dryers equipped with wet scrubbers to remove particulates, and rubber-lined chutes and screens to cut down on noise, made the construction process more complex and time consuming.*

Northwest Coal shortened the building time of its Western Slope preparation plant significantly by beginning construction before all of the necessary federal and state environmental permits were obtained. This plan involved a certain amount of risk. According to the A/E/C firms profiled by INFORM, the average time required to obtain permits is one and one-half years. In most cases, the construction of a preparation plant does not begin until all the necessary environmental permits are obtained.

However, Western Slope had contractual obligations with the plant's coal buyer, Central Illinois Light Company, and the plant had to be built as quickly as possible. Consequently, Allen & Garcia, Western Slope's A/E/C firm, designed and built the plant with careful attention to applicable environmental laws and regulations, and the company was able to begin operations within ten months.

---

*The Rum Creek plant's construction was also delayed by a miner's strike during construction when the construction workers refused to cross the picket lines.

Table F: PREPARATION PLANT SIZE, LEVEL, BUILDER AND CONSTRUCTION TIMES

Owner/ Plant	Size (tph)+	Level*	Builder	Time Required For Construction	Beginning of Commercial Operation
AEP Helper	1,100	4	Dravo, Inc.	NA	1977
AT Massey Marrowbone	NA	4	Dravo, Inc.	NA	10/79
Duquesne Warwick	900	3	Dravo, Inc.	Approximately 2 years	1974
Island Creek Hamilton	800	4	Lively Manufacturing & Equipment Co.	Coarse plant: 1½ years Fine plant: 7 months	Coarse plant: 1968 Fine plant: 1/79
Monterey No. 2	1,350	3	McNally-Pittsburg Manufacturing Co.	Approximately 3 years	7/77
Northwest Coal Western Slope	400	3	Allen & Garcia Co.	10 months	1/80
Peabody Randolph	1,676	4	Roberts & Schaefer Co.	NA	NA
PENELEC Homer City	960	5	Heyl & Patterson, Inc.	"Medium" circuit: + 2 years; Extensively-cleaned circuit: +3 years	"Medium" circuit: 12/77; Extensive: Fall, 1978
PP&L/PMC Greenwich	900	4	Coarse plant: Lively Mfg. & Equip. Co.; Fine plant: Roberts & Schaefer Co.	Coarse plant: over 1 year Fine plant: 2 years, 8 months	Coarse plant: 1971 Fine plant: 1/76
Pittston Rum Creek	500	4	McNally-Pittsburg	3½ years	3/80
R&F Warnock	800	4	Roberts & Schaefer	1 year, 9 months	8/78
TVA Paradise	1,607	4	Roberts & Schaefer	2 years, 11 months	6/81

*See Part III for a complete discussion of preparation and plant levels
+tph = tons per hour; cleaned coal output

When building a preparation plant, the "lead time" between the signing of a contract and the beginning of construction is usually between three and six months. There are exceptions, such as the Western Slope plant, which was built with only two months lead time. During this lead time, the design and engineering work is accomplished with the A/E/C firm in charge of the construction project.

Most of the preparation plant builders and owners profiled by INFORM reported that lead times will either remain the same or become shorter in the future. Roberts & Schaefer and McNally Pittsburg suggested that the increased use of computers, coupled with the ever increasing experience that A/E/C firms are developing in coal preparation plant projects, will shorten the time needed to design and engineer preparation plants.

Costs of Coal Cleaning

The cost of building, operating and maintaining a coal preparation plant varies, depending on the size of the plant, the amount and complexity of its equipment and the degree of automation. Inflation has driven up the cost of newer plants, and maintenance costs are often higher during the "start-up" phase of plants than they are after final operating adjustments are made and the personnel running the plants have become more familiar with them. Operation and maintenance (O&M) cost information is the most sensitive area of inquiry for preparation plant owners, and nine of these in INFORM's sample considered such information proprietary.

Cost information compiled by INFORM is presented in Table G. The figures, however, are not readily comparable. Dollar values have changed since many of the plants were built, and capital cost figures were not always furnished in the same year (for example, 1979 dollars). O&M costs are given in early estimates as the cost per ton of either raw or cleaned coal.

Nine of the 12 preparation plant owners supplied capital cost estimates for their plants. These ranged from $2.2 million (Northwest Coal/Western Slope) to $97.2 million (PENELEC/Homer City) in 1979 dollars. In most cases, the cost of building the plants was directly related to their size. For example, R&F's Warnock plant, with a level 4 cleaning ability using heavy-media equipment and processing 1,000 tons of coal per hour, was much more expensive ($22 million) than the

Table G: COSTS OF PREPARATION PLANTS

Owner/ Plant	Capital Costs	Operation and Maintenance Costs	Financing
AEP Helper	$10 million (1977 dollars)	NA	NA
AT Massey Marrowbone	NA	NA	NA
Duquesne Warwick	$15 million (1979 dollars)	NA	Sale of Duquesne Light Company stocks and bonds
Island Creek Hamilton	Coarse plant: NA Fine plant: $6 million (1979 dollars)	$1.00 to $1.50 per ton of raw coal processed	Internal cash flow from parent company (Occidental Petroleum); amortized over 20 years; investment tax credits given by government for plant construction
Monterey No. 2	$25 million (1977 dollars) includes front/ back loading equipment	NA	NA
Northwest Coal Western Slope	$2.2 million (1979 dollars)	NA	Capital funding from parent company (Northwest Energy Corporation); investment tax credits given by government for plant construction
Peabody Randolph	NA	NA	NA
PENELEC Homer City	$97.2 million (as of 12/81)	$13 million per year (estimate)	Bond sales and reinvestment of company profits
PP&L/PMC Greenwich	$13.7 million (original costs) $19.9 million (estimated costs in 1979 @8% increase per year)	$3.6 million (first six months of 1980; yearly estimate: $5.4 million; $2.75 per ton of cleaned coal	Leases/loans; initial financing: $9 million loans, $4 million leases; 20-25 year pay-back period; floating rate of interest on loans
Pittston Rum Creek	NA	NA	NA
R&F Warnock	$22 million (1978 dollars) (+ $8 million for front/back loading equipment and cleaned coal storage facilities)	NA	Capital funding from parent company (Shell Oil Company)
TVA Paradise	$44 million - preparation plant ($68 million - conveyors and auxillary equipment) (1981 dollars)	$1.39 per ton of cleaned coal	Electric power sales revenue; bond issue

Northwest Coal's Western Slope plant ($2.2 million), which cleans 400 tons of coal per hour only to level 3, using a jig.

Two plant owners (PENELEC and PP&L/PMC) provided detailed cost breakdowns that illustrate the various subcategories of preparation plant costs (see table H).

The $97.2 million capital cost of the PENELEC/Homer City plant was unusually high, not only because it is a large and complex plant, but also because of the extensive modifications that have been necessary to correct design errors. The plant's capital costs were originally projected to be about $55 million. (See Table H and Homer City profile for more information.)

PP&L/PMC also broke down the capital cost figures for their Greenwich plant. This preparation plant was built in two phases, with a coarse coal cleaning plant completed in 1971 and a fine coal cleaning

Table H:

Capital Cost Breakdown for PP&L/PMC-Greenwich Plant*

	Original Cost	Estimated Cost*
Buildings & Structures:		
(Coarse circuit) - Acquired in 1971:	$ 1,129,680	$ 2,090,959
(Fine circuit) - Acquired in 1976:	3,573,507	4,501,589
Total Cost:	$ 4,703,187	$ 6,592,548
Raw Bin & Screening Structure (1971):	$1,842,230	$3,409,839
Coal Refuse & Handling; Clean Coal Storage (1971):	$1,189,669	$2,201,994
Machinery, Labor, Materials (1976):	$5,797,000	$7,302,550
Raw Coal Storage System (1971):	$ 188,969	$ 349,264
Total Plant Assets (1971):	$ 4,350,548	$ 8,052,561
Total Plant Assets (1976):	9,370,507	11,804,140
Grand Total:	$13,721,055	$19,856,701

*adjusted to 1979 dollars @ 8% inflation increase per year

Capital Cost Breakdown for PENELEC/Homer City Plant
(December 1981)

High-Gravity Section: (Medium-Cleaned Coal)	$19,800,000
Low-Gravity Section: (Extensively-Cleaned Coal)	$32,300,000
Modifications:	$38,100,000
Truck Dump:	$ 7,000,000
Total:	$97,200,000

plant completed in 1976. Thus the estimated capital costs of the entire plant ($19.9 million) include the cost of constructing buildings and ancillary facilities (storage and refuse handling facilities), and buying plant equipment for each phase of construction. As can be seen in Table H, the fine coal cleaning plant (buildings and equipment) was more costly than the less sophisticated coarse cleaning plant, even after adjustment for inflation.

Operation and maintenance cost estimates were provided by only four plant owners in three different formats (yearly estimates, costs per ton of raw coal processed, and costs per ton of clean coal processed). The major components of these costs include labor and supervision, electric power, O&M supplies, lubricants, magnetite, reagents and water. Even though there is little basis for comparing O&M costs, PENELEC reported that their plant's $13 million per year O&M costs were high because of the high costs of starting up the preparation plant, and the design-related difficulties that led to more costly O&M.

Coal preparation plants are owned by three general categories of companies: utilities, oil/energy companies and "independent" coal companies. The utilities financed capital expenditures primarily by selling stocks and bonds (e.g., Duquesne Light Company), and by obtaining leases and loans from banks (e.g., PP&L/PMC). The oil/energy companies obtained plant financing from their internal cash flow, which in some cases included the parent company's monies (e.g., Island Creek and R&F Coal Companies). The independent coal companies did not respond to questions concerning the financing of preparation plants.

The only government incentive that companies purchasing preparation plants qualify for is the 10 percent investment tax credit for plant construction.* A number of plant owners believe that other government incentives should exist to aid potential preparation plant builders.

---

*Investors in preparation plants are not eligible for pollution control bonds (which have a lower interest rate than regular bonds) as are companies purchasing FGD units.

Chapter 3

Impact of Burning Cleaned Coal

Introduction to the Impact of Burning Cleaned Coal

By removing varying amounts of sulfur, ash and other impurities, the cleaning processes change the content and nature of many kinds of coal, and improve them as fuels to meet a variety of user needs. INFORM examined ten utilities burning coal that has been moderately to extensively cleaned. These utilities, as well as the seven coal companies interviewed, reported six areas of significant benefit which are summarized below and discussed in this section:

1. Although coal cleaning alone cannot enable a utility to meet emission standards for new boilers, it can remove enough sulfur to meet standards in some older boilers. For 7 of the 10 utilities, in 11 of their 16 power plants which were profiled, the use of a cleaned coal was sufficient to meet sulfur dioxide emissions standards.

2. Cleaning can yield a fuel which, used in combination with scrubbers, can insure that utilities that want to use high sulfur coal can consistently meet the most stringent sulfur dioxide emission standards. Two plants owned by two utilities in INFORM's sample will use cleaned coal for this purpose.*

---

*Two other utilities, which currently use cleaned coal with FGD, did not specifically state that they use cleaned coal to insure continuous compliance with emissions standards.

3. Eight utilities and seven coal companies noted one or more of the following benefits to scrubber operations from burning cleaned coal: 1) improved efficiency and availability; 2) less consumption of energy; 3) reduced waste disposal; and 4) lower operation, maintenance and capital costs.

4. All ten utilities stated that coal cleaning can benefit boiler operation in four ways: 1) improved efficiency; 2) reduced maintenance requirements; 3) greater availability; and 4) increased generating capacity.

5. Utilities cited mixed effects on ancillary power plant processes. While cleaning reduced pulverizer operation and waste handling requirements, in some cases it made particulate removal more difficult. In addition, increased moisture and coal dust caused handling difficulties.

6. According to all ten utilities, cleaning can result in lower shipping costs by reducing the amount of fuel to be transported. In INFORM's sample, for a given number of Btus, cleaned coal weighs 23 percent less than the same coal raw. Two coal companies reported that the reduced amount to be shipped can lessen pressures on the U.S. coal transportation system.

Under the current state of practice, we could, if the public and its elected representatives so chose, make more use of coal cleaning than we do. The President's Commission on Coal (1980) projected that U.S. utilities will consume 677,082,000 tons of coal in 1985. Assuming the average sulfur content of all U.S. coals to be 3.02 percent,[3] the coal used by utilities will contain 20,447, 876 tons of sulfur. If all this coal were to be cleaned to the average 25 percent of sulfur removal achieved in IN-FORM's sample, if utilities did not use coal whose raw sulfur content were higher than the present 3.02 percent average, and if EPA did not grant sulfur "credits" to utilities burning cleaned coal, then 5,111,969 fewer tons of sulfur would be burned in 1985.

Today, however, the quality of the nation's air is determined to a large extent by what the law requires. For the utilities in INFORM's sample, the impetus to clean coal to reduce its sulfur content is essentially a matter of complying with air quality standards. The extent to which their use of cleaned coal (with a lower sulfur content) increases or decreases depends on how clean we want our air to be and what we are willing to pay for it.

Meeting Air Quality Standards

HOW COAL CLEANING CAN HELP
MEET AIR QUALITY STANDARDS

Reducing Sulfur

All of the 16 power plants owned by the ten utilities profiled by IN-FORM burn cleaned coal. Ten plants (owned by seven utilities) burn a cleaned coal that would not meet emission standards if burned raw. (See Table A for a listing of these plants.) Of these plants, nine can meet emission standards using only cleaned coal, without the assistance of an FGD system. One plant and the newest unit of a second will begin using cleaned coal in combination with scrubbers to meet sulfur dioxide standards.* All seven utilities burning cleaned coal to help meet sulfur dioxide emission standards stated that generally, they set specifications for the characteristics (sulfur, ash, moisture and Btu) of the coal to be delivered to their boilers based upon boiler design and environmental constraints rather than merely stipulating "cleaned coal." If a coal supplier cannot meet the utility's specifications with its raw coal and wishes to be a supplier, it is forced to clean its coal. Likewise, if a utility's own coal reserves do not meet the specifications necessary for its boilers, it must clean its coal before burning it.

Decreasing Sulfur Variability

Four of the nine utilities interviewed (AECI, CP&L, NEES and TVA) emphasized the importance of coal cleaning as a way of making the sulfur dioxide concentration of the coal they used less variable; thus helping them meet emissions standards.† Fluctuations in the sulfur

---

*Two Duquesne plants in our sample also use scrubbers and burn varying percentages of cleaned coal. However, the company did not provide enough information to determine whether burning cleaned coal in these boilers helps meet sulfur dioxide emission standards.

†According to a 1980 study by Versar, Inc., coal cleaning can reduce the variability of the sulfur in a coal by 10 to 70 percent.[4]

content of coal can cause fluctuations in sulfur dioxide emissions, and may cause emissions to exceed standards intermittently. Thus, coal with a lower average sulfur content must be used to insure compliance. The Revised New Source Performance Standards (RNSPS) permit 30-day averaging on a rolling basis for monitoring emissions. Compliance with the old New Source Performance Standards (NSPS) is now based on a 3-hour test. However, EPA is considering switching the basis for compliance with this standard to a 30-day rolling average as well. Some state standards are more strict, requiring emissions to be averaged every 24 hours, or requiring continuous compliance. The shorter the averaging time, the less leeway there is for fluctuations in the sulfur dioxide emissions and in the sulfur content of the coal, and the more urgently the utilities need consistency in the feed coal to insure their compliance with air quality standards.

In Combination with Flue Gas Desulfurization

Two of the 16 power plants profiled that burn cleaned coal will use it in combination with FGD to meet emission standards. By 1982, Unit 3 at AECI's Thomas Hill station, and by 1984, TVA's Paradise power plant will have operating FGD systems and will continue to burn cleaned coal to help meet standards.

These two utilities stated that coal cleaning, because it reduces both the amount of sulfur in coal and the variability of the sulfur content, when used in combination with FGD, will enable their boilers to comply with sulfur dioxide regulations. Unit 3 of AECI's Thomas Hill Station, which will begin operating in 1982, must meet the NSPS which established a maximum emissions ceiling of 1.2 lb $SO_2/10^6$ Btu. If this unit burned raw, high sulfur Missouri coal, fluctuations in the sulfur content would at times cause the unit to exceed its sulfur dioxide emission limit even with scrubbers. Coal cleaning, by reducing the sulfur content of the coal 38 percent (from 11.55 to 7.18 lb $SO_2/10^6$ Btu) and making the sulfur content more uniform, enables the utility to meet its very lenient state standard while burning high-sulfur local coal.

Table A:
Cleaned Coal and SO_2 Standards

Power Plant, Location and Utility	SO_2 Standard lbs/10^6 Btu	Cleaned Coal Helps Meet SO_2 Standards	Use Scrubbers
Mountaineer New Haven, WV Appalachian Power Co., an AEP subsidiary	1.2	No	No
Tanner's Creek Greendale, IN Indiana & Michigan Electric Co., an AEP subsidiary	NA	NA	NA
New Madrid Units 1&2 Marston, MO Associated Electric Cooperative	10.00	Yes	No
Thomas Hill Unit 1 & 2 Unit 3 Moberly, MO Associated Electric Cooperative	9.5 1.2	Yes Yes	No Yes (will begin operating in 1982)
Roxboro Units 1,2 & 3 Hyco, NC Carolina Power & Light Co.	2.3	No	No
E.D. Edwards Unit #3 South of Peoria, IL Central Illinois Light Co.	1.8	No	No
Cheswick - Springdale, PA Elrama - Elrama, PA Phillips - Wireton, PA Duquesne Light Company	2.8 .6 .6	Yes NA NA	No Yes Yes
Brayton Points Units 1,2 & 3 Somerset, MA New England Electric System	*	Yes	No
Homer City Units 1 & 2 Unit 3 Homer City, PA Pennsylvania Electric Co.	3.7 1.2	Yes Yes	No No
Montour Washingtonville, PA Pennsylvania Power & Light- Pennsylvania Mines Corp.	4.0	Yes	No

*The Brayton Point Power Plant is subject to sulfur dioxide emission
regulations set by the State of Massachusetts--1.21 lb sulfur/10^6Btu
on a 30-day average, or 2.31 lb sulfur/10^6Btu during a 24-hour period.

Cleaned Coal and SO2 Standards (cont.)

Power Plant, Location and Utility	SO$_2$ Standard lbs/10^6 Btu	Cleaned Coal Helps Meet SO$_2$ Standards	Use Scrubbers
Gibson East Mount Carmel, IN Public Service Indiana	5.8	Yes	No
Colbert Mussel Shoals, AL Tennessee Valley Authority	4.0	Yes	No
Johnsonville Johnsonville, TN Tennessee Valley Authority	4.5	Yes	No
Paradise Drakesboro, KY Tennessee Valley Authority	5.2	Yes	Yes (Will begin operating in 1984)

Helping to Meet the Revised New Source Performance Standards

Coal cleaning can also help power plants now being built (all those on which construction was started after Sept. 12, 1978) which must meet the RNSPS requiring a 70 to 90 percent sulfur dioxide reduction. None of these newer boilers were profiled by INFORM. The RNSPS grants credit toward meeting the 90 percent reduction standard for the percentage of sulfur removed by coal cleaning. Thus, cleaned coal will require a lower level of sulfur removal by a scrubber to meet the overall 90 percent reduction requirement.

The Tennessee Valley Authority stated that the benefit to utilities of coal cleaning in meeting the RNSPS is greater than might be expected. There is significantly more than a percent for percent trade-off between coal cleaning and necessary scrubbing. Every percentage removed by coal cleaning decreases the efficiency needed by the FGD system by much less than one percent. For example, for a coal with 10 lb SO$_2$/10^6 Btu, the RNSPS would require a 90 percent overall reduction of SO$_2$ to 1.0 lb SO$_2$/10^6 Btu. If coal cleaning reduced potential emissions by 25 percent to 7.5 lb SO$_2$/10^6 Btu, then scrubbers would have to eliminate an additional 6.5 lb SO$_2$/10^6 Btu to lower sulfur dioxide emissions to the required 1.0 lb SO$_2$/10^6 Btu. This means scrubbers must operate at 87 rather than 90 percent reduction efficiency. The advantages of allowing even some of the flue gas to bypass the scrubber are discussed in the next section. Improved FGD Performance.

CLEANING COAL PROVIDES AN ALTERNATIVE TO SHIPPING WESTERN COAL AND FLUE GAS SCRUBBING IN MEETING AIR QUALITY STANDARDS

Coal cleaning is one of many strategies that utilities can use to meet sulfur dioxide standards for older boilers subject to State Implementation Plans and the NSPS. When utilities can use cleaned coal to meet their emission reduction requirements, they may be able to reduce or eliminate more expensive or more difficult alternatives; local coal can be used, avoiding the high costs and uncertainties of long distance coal transportation; and the need for FGD systems can be eliminated.

Shipping Low-sulfur Western Coal to Midwestern and Eastern States

Five utilities in our sample noted that cleaning coal prior to combustion allowed them to burn local coal and still comply with sulfur dioxide standards.* Three of these five utilities (AECI, CILCO and PSI) are located in the midwest where the sulfur content of the coal is relatively high. A representative from CILCO pointed out that coal that is mined in Illinois has been cleaned since 1950 to improve its quality, and that today washing this high-sulfur coal is even more important in light of the increasingly stringent sulfur dioxide removal requirements.

Two of the utilities in our sample however, do bring western coal east: CILCO brings Colorado coal to Unit 3 at its E.D. Edwards Station in Illinois, and AEP brings coal from Utah to its Tanner's Creek Station in Indiana. Central Illinois expressed concern over the long distance their coal has to be transported.

---

*As noted earlier, coal cleaning alone will never enable a new boiler (built after 1978) to meet sulfur dioxide standards that require a 70 to 90 percent sulfur reduction.

Six of the nine utilities interviewed said that in most cases, there are two disadvantages in bringing low-sulfur coals east to help meet sulfur dioxide emission standards. First, the Btu content of western coal is generally too low to assure efficient burning in most existing eastern boilers, which were generally designed to burn a higher Btu fuel. Carolina Power & Light's boilers, for example, are designed to burn a coal with 12,000 Btus per pound. The average range for the Btu content of western coal is from 8,000 to 10,000 Btus per pound. And second, the cost of transporting coal from the western to the eastern states is high. For example, in 1980 transporting coal by rail from Colorado to Illinois cost $22.44 per ton, while the average national cost of transporting a ton of coal by rail a distance of 300 miles was $8.04.[5]

Eliminating the Need for Scrubbers

Of the 12 power plants in our sample that now burn cleaned coal to meet sulfur dioxide emission standards, only one, PENELEC's Homer City power plant, burns cleaned coal to avoid having to use scrubbers. It must meet the NSPS for SO_2 of 1.2 lb $SO_2/10^6$ Btu. PENELEC's Homer City preparation plant was designed as the most cost effective means for the Homer City power plant to meet sulfur dioxide emission standards. Because it expects to obtain a very clean, low sulfur coal (0.84 percent) for Unit 3, PENELEC did not install scrubbers on this unit. Without extensive cleaning, the company would have to remove some of the sulfur dioxide when burning this coal by scrubbing to meet the NSPS standard.*

Improved FGD Performance

Three plants operated by two of the utilities studied by INFORM now use a combination of cleaned coal and lime or limestone scrubbers: the CILCO/Duck Creek Unit 1, and Duquesne Light's Elrama and Phillips stations. Scrubbers are now being constructed at three power plants owned by three other utilities studied by INFORM: the

---

*However, PENELEC's Homer City preparation has not yet cleaned its coal sufficiently to meet the 1.2 standard. For information on when and how PENELEC plans to improve its plant's capabilities, turn to the PENELEC profile.

AECI/Thomas Hill Unit 3, the PSI/Gibson Unit 5, and the TVA/Paradise plant. These stations will continue to burn cleaned coal when the scrubbers begin operating. All five of these utilities, as well as three other utilities and seven coal companies interviewed by INFORM, noted one or more of the following benefits to scrubber operations from using cleaned coal: 1) improved efficiency and availability; 2) reduced energy consumption; 3) reduced waste disposal; and 4) reduced operation, maintenance and capital investment needs.

Little information, however, is available to quantify the costs and benefits of using coal cleaning and FGD in combination. Of the two utilities in INFORM's sample that have combined cleaned coal with FGD, only Duquesne claimed to have data comparing the costs of scrubbing cleaned versus raw coal. However, Duquesne would not provide INFORM with this information. Moving beyond the sample companies in this report, INFORM reviewed three studies prepared in the past four years which sought to compare these overall costs, but their information is based on hypothetical power plants and is highly site specific.[6,7,8] Hence, the four advantages cited above can be discussed only generally.

Improved FGD Efficiency and Availability

The use of cleaned coal helps make scrubbers run more efficiently. Because 20 to 40 percent of the coal's sulfur has already been removed, less absorbent, such as lime or limestone, is needed. Burning cleaned coal also reduces "scaling" or the amount of impurities that adhere to and eventually corrode the walls of a scrubber.

Burning a cleaner and more uniform fuel reduces the wear and tear on a scrubber and may improve its "availability" or the amount of time it is available for use when called upon to operate. While TVA provided INFORM with no precise estimates, it claims, on the basis of an in-house study, that scrubbers work more of the time when cleaned coal is burned.

Greater availability of scrubbers is important to power plant operations. When scrubbers are not working, boilers often have to be shut down, particularly if they cannot meet sulfur dioxide emission standards. As boilers themselves (as of 1978) have an average availability of only 63 percent,[9] the more they can rely on scrubbers working, the more efficient the whole utility's operations will be. For utilities strapped for capacity, a high level of scrubber and boiler availability is especially critical.

Reduced Energy Consumption

Of the five utilities in INFORM's sample that use or will use a combination of cleaned coal and FGD, only TVA and Duquesne have quantified the impact this combination has on reducing flue-gas reheating requirements; however, they would not release this information. Three published studies have found that burning cleaned coal reduces the energy consumed by an FGD system. [10,11,12] (A 1978 study found that a limestone FGD system requires 2.5 to 6.1 percent of a unit's total energy output when 100 percent of the flue gas is treated.[13]) In fact, a lower flue-gas reheating requirement is one of coal cleaning's primary advantages to the most common "wet throwaway" scrubber systems (those that use water and do not recover their wastes). A 1980 PEDCo report found that the extent to which coal cleaning reduces the cost of running an FGD system depends primarily on the amount of flue gas that does not need to be reheated.[14] The amount of electrical energy not used by reheating can be sold to customers.

FGD systems that use liquids in the scrubbing process require flue-gas reheating. As the flue gas leaves the boiler, liquid sprays cool it from 300 °F to 124 °–130 °F. If not reheated, the cooled flue gas tends to condense as a weak acid on various parts of the exhaust system, causing corrosion, or it tends to settle in a cloud over the plant instead of dispersing into the upper atmosphere. To prevent this, scrubbed flue gas is reheated to about 175 ° before being sent up the stack.

By removing a significant amount of sulfur and ash before combustion, cleaning also reduces the scrubbing needed to meet sulfur dioxide emission standards, and thus, the amount of gas that must be reheated and the equipment involved in reheating. A 1979 EPA/TVA study and a 1980 report by PEDCo state that if 30 percent of the flue gas is sufficiently clean to bypass the scrubber, that gas can be used to warm the other 70 percent sufficiently to eliminate the need for reheating.[15,16] However, under the RNSPS, 90 percent of a coal's sulfur must be removed, making a 30 percent bypass impossible. The potential benefit of coal cleaning in this area is, therefore, more significant for power plants regulated under the less-stringent NSPS. However, even a partial bypass of flue gases, which coal cleaning can permit under the RNSPS, can reduce the energy requirements of a scrubber.

Reduced Waste Disposal Requirements
at the Power Plant

Because coal cleaning removes sulfur, ash and other impurities, less FGD sludge is generated, reducing the need for obtaining land and permits for disposal, transporting waste to a disposal site, stabilizing sludge or mixing it with flyash, and other activities involved in FGD waste disposal.

According to one study by the U.S. Department of Commerce, when moderately cleaned coal (level 3) is used in combination with lime scrubbing to meet the old NSPS of 1.2 lb $SO_2/10^6$ Btu, about 60 percent fewer total wastes (including ash, scrubber sludge, and sludge fixing agent) are generated at the power plant site than when the same raw coal is burned.* When coal which has been cleaned more extensively (level 4) is used in combination with lime scrubbing, the study found that over 70 percent less waste is generated. The waste to be disposed of when moderately cleaned coal was burned equaled 560,000 tons a year rather than 1,330,000 tons when raw coal was burned. For moderate cleaning, however, 500,000 tons of coal refuse must be disposed of at the preparation plant site. Because less coal, (less coal is needed because cleaning concentrates the number of Btus per pound) less lime, and less sludge fixing agent are used when cleaning and scrubbing are combined, the amount of waste generated at the preparation plant and at the power plant together is still less than the waste generated at the power plant using lime scrubbing and raw coal (1,060,000 tons compared to 1,330,000 tons). (Turn to Table B for comparisons.)

A reduction in the amount of waste generated is especially important for utilities converting power plants to coal that have a limited amount of space available for the contingencies of coal burning, storage and waste disposal and also for utilities building a power plant in populated areas where space for disposal is limited and/or expensive.

---

*This study assumed fuel requirements and waste products for new 1,000 Mw power plants burning average Northern Appalachian coal (11,000 Btu per lb, 3 percent sulfur, 12 percent ash).[17]

Table B

GROSS SURFACE ENVIRONMENTAL CONSIDERATIONS FOR EMISSION CONTROL TECHNOLOGIES*

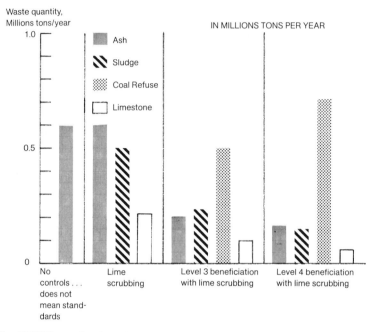

*New 1000 MW generating plant using 3.0% sulfur Northern Appalachian coal and meeting New Source Performance Standards   *(Table reproduced courtesy of EPRI)*

Reduced Operation, Maintenance
and Capital Investment Needs

Increased efficiency, reduced energy consumption, and the generation of less waste all can make existing scrubbers operating with cleaned
coal easier and less expensive to operate and maintain than those
operating on boilers burning raw coal. All of these factors will also have
an impact on the FGD systems yet to be built. By increasing FGD efficiency and availability, coal cleaning can eliminate the need to over-
design a scrubber's capacity to remove sulfur dioxide. In addition, by
reducing sludge, cleaning reduces the need for waste handling and
thickening capacity. Finally, because cleaned coal burns slightly more
efficiently and produces slightly less flue gas, a smaller sulfur absorber
and fewer recirculation pumps need to be installed.

Improved Boiler Performance

Representatives of all ten utilities profiled by INFORM cited
potential benefits to boiler operation from burning cleaned coal. These
benefits all result from the decrease in ash, sulfur and other non-
combustibles in the fuel. The four general categories of benefits cited
were: 1) improved efficiency; 2) reduced maintenance requirements; 3)
improved boiler "availability"* and 4) increased electrical generating
capacity. However, only three of the utilities profiled (AEP,
PENELEC, and TVA) have quantified any of these benefits. (IN-
FORM reviewed all recent industry and government studies dealing
with the impact of cleaned coal on boiler performance and found that
they contain little quantitative data.) All the utilities emphasized the
importance of burning the coal that boilers were designed to burn and
suggested that more and more, this will require the use of cleaned coal
as a result of the decreasing quality of coal.

The utilities INFORM interviewed offered two explanations of
why these benefits have not been quantified: because there is very little
data available to analyze, and because utilities, until recently, have
often not kept records tracing the quality of coal fed to their boilers and
of these four categories of boiler performance. In addition, in order to

---

*The amount of time that a boiler is available to operate, which determines
 how much electricity it can generate.

determine the effect of cleaned coal, at least 30 different aspects of boiler performance must be studied by the utility.

Efficiency

Decreasing a coal's ash and mineral content (particularly iron, sodium and calcium) generally reduces slagging, fouling and tube corrosion in a boiler using the fuel. As a result, the boiler is subject to fewer outages which, in turn, reduces boiler maintenance requirements, improves boiler availability and efficiency, and sometimes increases generating capacity. Being more consistent in quality than raw coal, cleaned coal also makes the boiler more easily adjusted. Raw coal can cause unwelcome surprises for a boiler operator as particles of coal may have a higher than usual sulfur, ash or moisture content, or a lower than usual Btu content. If the constituents of a coal are very variable, at any particular moment during combustion the boiler can be adjusted incorrectly. With cleaned coal, the amount of each of these constituents is more constant and boilers are more often adjusted properly for the coal they are burning, and thus overall boiler efficiency is improved.

Although all ten utilities cited cleaning as a means of improving the fuel, two utilities interviewed, CILCO and PENELEC, noted that cleaning can, in a few cases, produce a boiler fuel that will cause more boiler problems than raw coal. Central Illinois Light Company demands cleaning for its coal from Utah to reduce ash and to produce a more consistent quality coal. However, it does not stipulate cleaning for its poor quality Montana coal, as CILCO believes that cleaning this coal would decrease boiler efficiency by adding additional moisture to the coal and concentrating its already high sodium content, which could increase slagging and fouling.

PENELEC had Babcock and Wilcox, the Australian Coal Research Group, and Pennsylvania State University study whether the combustion of intensively cleaned low sulfur coal would lead to slagging and fouling at its Homer City Unit 3. The study results indicated that it would not. However, in some instances cleaning can cause problems in the boiler by changing the relative concentration of mineral materials in a coal.

Increased Availability and Capacity

Four of the ten utilities profiled (AECI, AEP, PENELEC and TVA) noted specific cases in which burning cleaned coal enabled a power plant to generate more electricity than would have been produced if raw coal was used, by increasing the availability of the generating units. Adding generating capacity to a utility which is strapped for capacity reduces the need for the utility to buy power from outside its electrical grid, or to build a new generating capacity to meet customer demands.

These benefits have been especially important in the past ten years, because the costs of new electrical capacity have skyrocketed, while boiler availability has deteriorated. According to AEP, over the past ten years the capital cost per kilowatt capacity of a generating unit has risen from approximately $100 to $600 without scrubbers.[18] Adding scrubbers, which are now required on all new units, increases the cost from $750 to $800 per kilowatt.[19] AECI states that rising fuel costs, high interest rates and increasing government regulation are making its new unit (to be completed in 1982) cost $730 per net kilowatt of capacity. By contrast, the original units at this power plant completed in 1966 and 1969, cost $137 per net kilowatt of capacity.

Not only have the costs of installing new generating capacity increased dramatically in recent years, but the national average of boiler availability of generating units has been deteriorating. From 1968 to 1978 boiler availability dropped from 78 to 63 percent.[20] This decrease in availability is partially attributed to a decrease in the average Btu content of coal delivered to utilities (from 11,700 to 10,600 Btus per pound) which can potentially decrease the number of kilowatts generated by a unit.[21]

Of the four utilities citing improved boiler availability from burning cleaned coal, two mentioned the resulting increased generating capacity, but were unable to document the percentage increase. The boiler tubes in AECI's New Madrid Units 1 and 2 became partially plugged when raw coal from southwestern Illinois was burned because of the large percentage of fire clay in the coal, even though the boiler was designed to burn this coal. As a result, the amount of electricity which the units could generate was reduced, occasionally forcing AECI to purchase higher-priced emergency electricity from other utilities in order to meet demand. AECI indicated that burning cleaned coal in Units 1 and 2 at New Madrid enabled the utility to meet present demands for electricity and to increase the electrical output of the plant.

TVA's Paradise power plant was plagued with premature break-downs and excessive fly ash in the boiler when burning high-sulfur coal. TVA responded to these problems by constructing the Paradise preparation plant, which will lower the sulfur and ash content of the coal burned at the Paradise power plant (see TVA profile).

In the third case, PENELEC stated that by a combination of concentrating the Btu content and decreasing the ash and non-combustibles in the fuel, coal cleaning improved boiler efficiency and availability, increasing generating capacity. PENELEC had encountered problems with boiler deposition and efficiency due to slagging and fouling at the Homer City Station. To improve boiler availability, first the company slowed the rate at which coal was fed into the boilers. This change resulted in less fouling, slagging and corrosion, but lowered the electrical output of the station by 70 to 80 Mw. When PENELEC switched to burning cleaned coal at this station, the average boiler rating increased by about 25 Mw, partially offsetting the effects of decreasing the amount of fuel fed to the boilers.

A study conducted by AEP between 1971 and 1975 of two identical generating units (identical design and constructed around the same time) burning a high quality and a lower quality (combined ash and sulfur: 15.2 versus 19.7 percent; heat content: 11,810 versus 11,460 Btu per lb; slagging index: low to medium versus very high; and ash fusion temperature: more than 2700 °F versus less than 2500 °F) coal,[22] also concluded that burning cleaned coal improved the availability of a generating unit. The availability of the unit burning the higher quality coal was 89.25 percent. The second unit burning an inferior coal showed a much lower availability of only 77.45 percent. This added approximately 12 percent to the "capacity factor"† of the first boiler by reducing boiler outages caused by slagging and fouling. AEP concluded from its study that there are "economic and operational justifications for burning a high Btu, consistent quality coal which, considering today's mining techniques, can only be achieved through the washing of the coal in a comprehensive preparation."[23] In a paper given by Gerald Blackmore, Executive Vice President of Fuel Supply at AEP in 1980,[24] AEP states that improved boiler availability is critical because if 10 percent of AEP's coal-fired capacity of 14,500 megawatts was lost due to

---

*The amount of electricity a boiler can generate over a given period of time.

coal-related outages (or a loss of 1,455 megawatts), making up this loss in generating capacity by the financing and construction of new facilities would cost in excess of $1 billion.[25]

According to AEP, in most cases cleaning coal to minimize coal-related boiler outages is a less expensive way of meeting electricity demand than constructing new generating capacity. Approximately 17 of the 40 million tons of coal purchased by AEP in 1980 needed to be cleaned to bring the coal's quality up to the specifications for the boiler. More than $14 per ton could be spent for cleaning this coal, according to the company, before the $250 million annual cost of new plant investment would be offset (this does not include savings for utilities in boiler operation and maintenance).[26] The higher cost of cleaned coal over raw coal is much less than $14 per ton, according to the companies in our sample that provided cost figures (Peabody and PENELEC). (Turn to the next section on Costs for quantitative data. For more information on AEP, turn to the AEP profile.)

Mixed Effects on Ancillary Systems of Generating Plants

Although few companies would use coal cleaning merely to improve ash handling and pulverizer operation, one or both of these were cited as additional benefits of cleaned coal by eight of the ten companies profiled by INFORM. Although coal cleaning also has an impact on coal handling and flue-gas particulate removal, these areas of coal cleaning received mixed reviews from the utilities (see table C). While cleaned coal has less ash and produces fewer particulates when burned, reducing the sulfur in coal may cause problems with the performance of electrostatic precipitators (ESPs) in removing particulates. Only two utilities felt that the advantages outweighed the disadvantages in this area.

Similarly, while two utilities reported that coal handling problems were reduced because there was less coal to handle, five companies felt that this advantage was more than offset by the added moisture in clean coal* and by the increases in very fine coal particles resulting in coal dust. Three reported no net effect.

---

*Although thermal drying is energy intensive and costly, it can reduce the moisture that can be added by coal cleaning.

Table C:
Effects of Coal Cleaning on Ancillary Systems of Generating Plants

	I	II	III	IV
Utility	Pulverizer	Ash Handling and Disposal	Coal Handling	Particulate Removal
American Electric Power Co.	NA	NA	NA	NA
Associated Electric Cooperative	NA*	+	+	+
Carolina Power & Light Co.	0	+	0	0
Central Illinois Light Co.	+	+	-	+
Duquesne Light Co.	0	+	0	0
New England Electric System	+	0	+	0
Pennsylvania Electric Co.	+	+	-	0
Pennsylvania Power & Light - Pennsylvania Mines Corp.	+	+	-	
Public Service Indiana	0	+	-	0
Tennessee Valley Authority	NA*	+	-	+

NA - Not Available   - - Negative effect noted

0   - No effect noted   * No pulverizers

+   - Positive effect noted

Ash Disposal

Coal cleaning decreases the amount of ash that a utility must remove from its boilers and dispose of at the power plant site. Consequently, there is less waste handling and treatment, less maintenance of waste handling systems, and waste disposal sites can be used longer. Although seven of the ten utilities profiled cited advantages in waste handling and disposal as significant benefits of coal cleaning, none provided figures to quantify the savings.

Pulverizer Operation

Cleaned coal is easier to pulverize than raw coal. By removing the harder and more abrasive coal constituents, ash and pyrite, and by partly crushing the coal, coal cleaning provides a fuel that reduces pulverizer wear and energy consumption. For example, the size of the cleaned coal pieces delivered to Central Illinois Light Company average no more than 2 inches in diameter, while raw coal pieces can be 4 to 5 inches in diameter. Similarly, 50 percent of the cleaned coal pieces delivered to New England Electric System are less than one inch in diameter.

Coal cleaning also concentrates the Btu value of the coal, so that less coal has to be burned to achieve a required energy output. For boilers that use pulverizers, 9 out of 10 in INFORM's sample, there is less coal to be crushed and therefore less energy used for this step of energy production, and reduced wear on the pulverizer.

Coal cleaning also increases the Btu capacity of a boiler, without increasing its coal crushing requirements. Hence, in some cases, new pulverizers are not needed when increasing energy output. This is especially important for utilities that must meet increased demands for power, that must burn a low Btu coal, or that encounter space restrictions which make building a new pulverizer impossible. As in ash disposal, however, the companies did not quantify these savings.

Particulate Removal

In INFORM's sample, two utilities, Associated Electric Cooperative and Central Illinois Light Company, found that cleaned coal was an advantage to particulate removal, while one reported that coal cleaning had a negative effect. For the other seven, cleaned coal had no significant net effect. Because most cleaned coal contains less ash, it produces fewer particulates when burned, reducing the removal requirements of ESPs needed for compliance with a given particulate emission regulation. Two studies (Buder and PEDCo)[27,28] found that the required size of an ESP at power plants where cleaned coal is burned may be lower by 5 to 25 percent of the size required where raw coal is burned. These capital savings apply, however, only to new plants where reductions in ESP size are made at the design stage. (For the cost savings given by these studies, turn to the Cost Section of the Findings.)

Coal cleaning can also impair the particulate removal efficiency of ESPs, as the susceptibility of particulates to removal decreases as a coal's sulfur content decreases. Thus, as coal cleaning reduces the sulfur content of a coal, the efficiency of the ESP is reduced as well. However, Central Illinois Light Company stated that problems with ESP operation can occur when burning any low sulfur coal (less than 2.5 percent), not just when cleaned coal is used.

Methods of improving the operating efficiency of an ESP when low sulfur coal is burned were reported by some of the utilities in IN-FORM's sample. Central Illinois' E.D. Edwards Station and Pennsylvania Power & Light's Montour Station condition their flue gas with SO_3 to enhance particulate collection. The baghouse is another particulate collection device which is as effective as an ESP in removing particulates, but whose performance is not impaired by physical coal cleaning as the removal mechanism operates according to properties of coal which are not changed by cleaning. However, Central Illinois noted that baghouses require much more space than ESPs, limiting their usefulness as an alternative for existing units using ESPs. (All the power plants profiled in INFORM's sample use ESPs, no baghouses are used to remove particulates.)

Coal Handling and Storage

Five of the utilities interviewed by INFORM stated that cleaning coal increased problems with coal handling and storage at a power plant. Two reported a positive effect and three no net effect on this part of utility operations. Cleaned coal takes up less room per Btu than raw coal. However, the utilities reported problems resulting from the finer-sized coal particles produced by coal cleaning and from the added moisture content of the coal: the finer particles sometimes stick to chutes and belts and cause increased dust emissions, and the increased moisture can cause coal particles to freeze together in winter (if the coal is not thermally dried).

Transportation: Lightening the Load

By eliminating the wastes that account for 20 to 40 percent of a raw coal's mass, cleaning reduces the weight that has to be transported for an equivalent number of Btus. As a result, shipping expenses are reduced, as are the pressures on the transportation system. As more coal is produced and transported over long distances, these benefits should become even more significant. However, who actually benefits and by how much is very difficult to calculate.

Reduced Shipping Expenses

Transporting coal, especially by rail, can be a major cost factor in the price of delivered coal. In 1979, railroad rates accounted for 26 percent of the cost of delivered coal; the average price per ton was $31.65 and the cost of moving it by rail was $8.15 per ton.[29] While using barges, trucks and slurry pipelines cost slightly less, these means are not generally as available as railroads.

Cleaning coal before shipping can reduce these costs. If the coal delivered weighs less and contains the same number of Btus, a saving in diesel fuel for the shipper is always realized. Part or all of this saving is passed on to the utility. Less coal also means that fewer deliveries need to be made or fewer vessels will be required for each delivery, and that loading and unloading the coal will take a shorter time. Since costs are based on dollar per ton moved, the amount saved should correspond to the number of tons of coal refuse removed by cleaning.

Ten of the preparation plants profiled by INFORM eliminate an average of 27 percent of a raw coal's impurities, in the process losing about 6 percent of the Btus. Therefore, for the same number of Btus, the cleaned coal weighs about 22 percent less than the raw coal.

Cost savings, however, are not necessarily proportionally lower as the weight is reduced. In fact, the cost of moving a load of cleaned coal is so dependent on site-specific factors that any generalizations are risky. While all of the utilities interviewed by INFORM agreed that coal cleaning reduces shipping expenses, neither these utilities nor the studies INFORM consulted, provided any data upon which generalizations could be based.*

Three variables have a particularly critical influence on the extent to which coal cleaning can reduce shipping costs: the percentage of raw

---

*The 1980 *President's Commission on Coal* has predicted that by 1985, 677,082,000 tons of coal will be used by U.S. utilities. If all of this coal were cleaned to the level of 23 percent weight reduction in INFORM's sample, 155,728,860 tons of that amount would be reduced.

Table D: CLEANED COAL AND TRANSPORTATION NEEDS

Utility	Power Plant and Location	Coal Supplier	Preparation Plant and Location
Appalachian Power Co., an AEP subsidiary	Mountaineer New Haven, WV	Pittston	Rum Creek Logan County, WV
Indiana & Michigan Electric Co., an AEP subsidiary	Tanner's Creek Greendale, IN	Price River Coal Co., an AEP subsidiary	Price Mine No. 3, Helper, UT
Associated Electric Cooperative	New Madrid Marston, MO	Peabody Coal Co.	Randolph Marissa, IL
Associated Electric Cooperative	Thomas Hill	Associated Elec. Coop.	Prairie Hill, MO
Carolina Power & Light Co.	Roxboro Hyco, NC	Marrowbone Development Corp., a subsidiary of A.T. Massey	Marrowbone Naugatuck, WV
Central Illinois Light Co.	E.D. Edwards South of Peoria, IL	Western Slope Carbon, Inc., a subsidiary of Northwest Coal Corp.	Hawk's Nest East & West - Paonia, CO
Duquesne Light Co.	Cheswick-Springdale, PA Elrama-Elrama, PA Phillips-Wireton, PA	Duquesne Light Co.	Warwick Greensboro, PA
New England Electric System	Brayton Point Somerset, MA	Many suppliers and preparation plants	
Pennsylvania Electric Co.	Homer City Homer City, PA	Pennsylvania Electric Co.	Homer City Homer City, PA
Pennsylvania Power & Light	Montour Washingtonville, PA	Pennsylvania Power & Light Pennsylvania Mines Corp.	Greenwich Collieries Ebensburg, PA
Public Service Indiana	Gibson East Mount Carmel, IN	Monterey Coal Co.	Monterey No. 2 Albers, IL
Tennessee Valley Authority	Colbert Mussel Shoals, AL	R & F Coal Co.	R & F Warnock, OH
Tennessee Valley Authority	Johnsonville Johnsonville, IN	Island Creek Coal Co.	Hamilton No. 1 Morganfield, KY
Tennessee Valley Authority	Paradise Drakesboro, KY	Tennessee Valley Authority	Paradise Drakesboro, KY

coal which ends up as clean coal (yield) and the actual increase in the number of Btus per ton, the rate structure of the shipping company, and the distance the coal is to be transported. The first variable, the yield and the Btu concentration in a ton of coal, are relatively straightforward and rely primarily on the type and intensity of the cleaning process, the amount of material discarded as refuse, and the percentage of Btus lost in cleaning (from 4.5 to 10 percent in INFORM's sample of preparation plants).

The second variable, the rate structure of the shipping company, tends to be highly erratic and often very complicated. Each rate is individually negotiated. In addition to the number of tons to be moved and the distance to be travelled, several other factors have an impact on the cost of an individual shipment: if railroads are to be used, the type and size of the train; whether the utility or the transport company owns the railroad cars; and the amount of time it takes to load and unload the coal. Adding or removing a carload can also affect the rate charged—as the amount to be transported increases, costs per ton generally decrease.

Two of the utilities interviewed by INFORM even suggested that transport companies will charge more for cleaned coal than they do for raw, but a Norfolk and Western official categorically denied this.

Finally, if chemically cleaned coal, which is crushed to a size smaller than 3/8 inch, becomes commercially available, transit costs can increase if the fine coal dust containing Btus blows off the rail cars.* Although handling this coal can add to shipping costs, a 1979 article by the Department of Energy concluded that coal size will at most minimally affect rail prices.[30]

The third variable, the distance the coal is transported, will increase in significance if more western coal is mined and transported to midwestern, eastern and European markets. In 1975, 111 million tons of coal were produced west of the Mississippi. By 1979, this figure had already grown to 215 million tons, or one-quarter of the national market.[31] By the year 1990, the Department of Energy projects that western coal production will reach 675 million tons a year. The more miles that coal must be transported, the greater the cost, and thus the greater the savings which may result from coal cleaning. For example,

---

*According to an EPRI official, losses from 1 to 2 percent of the Btu content of a trainload of physically cleaned coal now occur.

Table E: LOGISTICS OF CLEANED COAL

Power Plant	Distance Between Coal Source & Power Plant (Miles)	Mode of Transport	Size of Cleaned Coal Particles	Transport Cost Per Ton	Weight Reduction in Cleaned Coal From Raw(%)	Btu Recovery in Cleaned Coal(%)	Reduction in Shipping Weight Without Btu Loss
Mountaineer	NA	barge	1½" to 0	NA	40	NA	--
Tanner's Creek	NA	NA	NA	NA	NA	NA	NA
New Madrid	220	barge	1½" to 0	$2.26 (1980 avg.)	24	NA	--
Thomas Hill	on-site	conveyor	2" to 0	--	--	--	--
Roxboro	NA	NA	NA	NA	NA	NA	NA
E.D. Edwards	1,339	unit train	2" to 0	$22.44	20 to 25	+90	14
Cheswick	less than 100	barge	5" to 0	.80	35	96	32.3
Elrama	less than 100	barge	5" to 0	.80	35	96	32.3
Phillips	less than 100	barge	5" to 0	.80	35	96	32.3
Brayton Point	800 to 900	rail & ship	2" to 0	--	--	--	--
Homer City	on-site	conveyor belt (cleaning plant on site)	1" to 100 mesh	--	20	95.4	16.4
Montour	196	unit train	2" to 28 mesh	$3.60	28	96.1	25.1
Gibson	117	unit train	2" to 0	NA	30 to 35	+90	25.0
Colbert	720	barge	2" to 0	$10.97	23	*	--
Johnsonville	210	barge	NA	$2.29	30	96.5	27.5
Paradise	on-site	conveyor belt (cleaning plant on site)	3" to 28 mesh	--	15	93	20.4

*Preparation plant will not begin operation until March of 1982.

a 1978 Gibbs and Hill study[32] found that using trains and barges to ship low sulfur Utah bituminous coal would be a "waste of both money and energy" without prior cleaning. By eliminating 12 percent of the yearly tonnage, Gibbs and Hill estimated that coal cleaning would result in savings substantial enough to fully pay for itself.

At present, the average shipping distance is 300 miles,[33] however, two of the utilities in INFORM's sample, Central Illinois Light Company (CILCO) and American Electric Power (AEP), receive coal from sources over 1,000 miles away. Central Illinois Light has a power plant, the E.D. Edwards station, south of Peoria, Illinois, which receives coal from the Western Slope Carbon preparation plant, 1,339 miles away in Somerset, Colorado. In December 1979, transporting one ton of this coal by rail cost $20.78, compared to the national average of $8.15 per ton. A 73-car "unit train" carrying approximately 10,000 tons of coal, makes about one trip a week to the E.D. Edwards staton. To get to Illinois, the train must use the tracks of three different railroads and cross the Rocky Mountains at 11,000 feet through Colorado's Tennessee Pass. The 5 engines pulling and 9 pushing the train consume 50,000 to 60,000 gallons of fuel for each round trip.

Central Illinois Light has not calculated the savings it gains by transporting cleaned coal, but they should be considerable. Assuming the cost of transporting a *volume* of raw coal and the *lesser volume* of cleaned coal to CILCO's E.D. Edwards Station to be the same, the utility theoretically saves $1,570,800 per year when transporting cleaned coal cleaned from the mine site (weighed against the extra cost of cleaned coal).* The Western Slope Carbon preparation plant reduces the raw coal's weight by 20 to 25 percent. For the number of Btus required, about 12 percent less coal is shipped clean than if it were raw. Without prior cleaning, 10 more cars would be needed and between 7,000 and 8,400 additional gallons of diesel fuel would be consumed each trip.

---

*This cost saving is based on the fact that 11,400 tons of raw coal, as opposed to 10,000 tons of cleaned coal, would to be shipped for the required number of Btus per delivery; that coal deliveries are made to the E.D. Edwards plant 50 weeks of the year; and that the 1980 cost of transporting a ton of coal this distance, given by CILCO, is $22.44.

Reduced Pressure on the Transportation System

Eliminating impurities before coal is shipped should also reduce some of the strain on the already overburdened transportation system, a system whose inadequacies, according to the World Coal study, will become increasingly apparent and an increasing liability, with the dramatic rise in U.S. coal production and export projected in the next two decades. In 1980, the United States exported 89.9 million tons of coal. By the year 2000, it is expected to export between 138 and 386 million tons. The World Coal Study concludes that the U.S. cannot handle this increase unless its ability to move coal across and out of the country improves.

All of the methods of transporting coal across the country have problems. Financial strains over the past decade have prevented railroads from investing the funds necessary to maintain their rolling stock (cars, engines, and rail beds), and if more coal is to be moved, rail bed repairs are essential. The amount of coal which can be carried by barge over inland waterways is also limited by the number and capacity of locks on any given river, but lock replacement is lagging behind traffic growth. Deteriorating roads are, many claim, stymieing coal production, especially in Appalachia.[34] And before more slurry pipelines can be built to transport coal, legal entanglements between pipeline investors and railroads for the right of "eminent domain" must be resolved. Coal cleaning cannot solve the problem of limited domestic capabilities, but by reducing the tonnage of coal at the mine by as much as 40 percent, it can lessen the pressure on the transportation network.

Port congestion, already a problem, according to many coal companies, can also be eased by cleaning coal (although this will mean that the United States, not the importing country, must deal with the wastes generated in the cleaning process). Ships in Hampton Roads, Virginia, and Baltimore, Maryland, must wait from three to six weeks in port for their loading turn. This wait can be costly. Claire C. Chasnov of the Association Techniques de L'Importation Charbonniere, which expects to buy 11 or 12 million tons of coal from the United States in 1981, estimated that loading delays have cost from $15,000 to $18,000 per ship per day, adding about $15, or 30 percent, to the price of a ton of coal (which in 1981 sold for about $45 or $50 a ton).[35]

Port congestion can also add to a coal company's costs by holding up rail cars, notes a Pittston spokesman, as companies are forced to stockpile the coal in their yards, rather than load it directly on the cars. Foreign customers warn that such delays and added costs may cause the United States to lose long-term contracts, particularly from the European power companies now switching from oil to coal.

The bulk of the increase in exports is expected to be steam coal used by power plants to generate electricity. Unlike metallurgical coal which accounted for two-thirds of the 1980 coal exports, and which is used to make steel, steam coal is not necessarily cleaned. Cleaning steam coal should, therefore, play a particularly important role by reducing the tonnage of coal to be exported.

Cost and Benefits of Burning Cleaned Coal

The data on the costs and benefits of coal cleaning are based primarily on the ten studies consulted by INFORM (listed in the chart below). Six of these studies provide estimates of both cleaning costs, and of the benefits of burning cleaned coal to different components of power generation (FGD, boilers, ancillary processes and transportation). Two studies look only at cleaning costs, and two look only at the user benefits of burning cleaned coal.

For the most part, the conclusions of these studies support INFORM's findings that the use of cleaned coal does save utilities money. One study by PEDCo (1980), looked at the whole range of costs and benefits investigated by INFORM. Using data from Hoffman-Muntner (1978), which was the only study to examine commercially operating preparation plants, PEDCo found the typical fuel cost of cleaning to be an average of $4.85 per ton. The same study found that a typical utility would save approximately $7.20 per ton of fuel if cleaned coal was used. Cleaning can thus yield a net gain of $2.35 per ton. Other studies, which had a more narrow focus, also generally conclude that several of the benefits of using cleaned coal result in savings for utilities — savings which should also benefit electric power consumers.

Author and Title	Costs/Savings	Basis of Estimates
Cole, Randy M./TVA "Economics of Coal Cleaning and Flue Gas Desulfurization with Revised NSPS for Utility Boilers" (April 1979).	Costs and savings	Hypothetical cases

Hoffman-Muntner Corporation Hoffman and Holt: *Engineering/Economic Analyses of Coal Preparation with SO₂ Clean-up Processes for Keeping Higher Sulfur Coals in the Energy Market* (November 1976).	Both costs and some benefits	Hypothetical cases
Hoffman-Muntner Corporation Holt, Elmer Jr.: *An Engineering/Economic Analysis of Coal Preparation Plant Operation & Cost* (July 1978).	Capital and O & M costs of cleaning	Eight real plants
PEDCo *Cost Benefits Associated with the Use of Physically Cleaned Coal* (March 1980).	Both costs and savings (One of first efforts to quantify all benefits associated with use of cleaned coal)	Cleaning costs based on H-M 1978 data; benefits based on range of studies
Pennsylvania Electric Company "Estimated Cost Comparison, FGD vs. MCCS" (November 1977).	Cost of cleaning (vs. FGD or FGD & PCC to meet standard)	Estimates for a plant to be built given by FGD suppliers and preparation plant manufacturers
Gibbs & Hills Phillips, Peter J.: *"Coal Preparation for Combustion and Conversion"* (May 1978).	Both costs and some savings	Costs based on co. experience; benefits based on literature review
EPA/TVA Tarkington, T.W., etc.: *Evaluation of Physical/Chemical Coal Cleaning and Flue Gas Desulfurization* (July 1979).	Cost of 3 levels PCC and 3 levels CCC, and benefits to FGD capital and operating costs only	Generic cases
Hoffman-Muntner Corporation Holt, Elmer C. Jr.: *Effect of Coal Quality on Maintenance Costs at Utility Plants* (June 1980).	Benefits to maintenance costs at utility plants	Actual maintenance costs of selected portions of five TVA plants
Electric Power Research Institute Vivenzio, T.A.: *Impact of Cleaned Coal on Power Plant Performance and Reliability* (April 1980).	Benefits from improved power plant performance and reliability	Examination existing raw vs. cc operating data of plants with experience
Buder and Clifford "The Effects of Coal Cleaning on Power Generation Economics" (April 1979).	Both costs and benefits	Seven hypothetical case studies

However, two factors make it very difficult to quantify the costs and economic benefits associated with coal cleaning. First, little information is available from companies cleaning coal or from utilities using cleaned coal. Preparation plant owners hesitate to reveal information on costs and income that they consider "proprietary," while most utilities have no precise data on the expense they have incurred from buying cleaned coal or the savings accrued from its use. Second, the information that does exist is highly site-specific, and generalizations must be made with caution.

Even taken together, the studies consulted by INFORM do not give a complete or precise picture of the economics of coal cleaning. Most of their estimates are calculated from figures based on hypothetical preparation or power plants, or on the actual benefits derived from burning coal of higher quality, rather than a coal that has specifically been cleaned. No study provided figures for the additional price that a preparation plant will charge for cleaned coal. Finally, many of the studies use different assumptions in measuring costs and benefits, making their conclusions difficult to compare.

Until more complete and reliable information is available on preparation costs and on the extra amounts charged for cleaned coal, and until the often subtle benefits of using cleaned coal can be translated into quantifiable data, coal cleaning's full economic value can only be estimated.

COSTS OF CLEANING

The cost of cleaning coal is determined by a number of factors that vary significantly from site to site: the capacity of the preparation plant and its mode of operation; the type of coal being washed; the degree of cleaning; and the local costs of resources such as water, electricity and labor. The cost of cleaned coal to the user depends on these factors and on the profit a preparation plant owner takes. The variability in these factors makes it difficult to compare the costs of cleaned coal from different preparation plants.

Of the 12 companies owning preparation plants interviewed by INFORM, only two would divulge information on cleaning costs. PENELEC told INFORM that moderately cleaning coal at its Homer City level 3 circuit costs $2 per ton; extensive cleaning costs three times as much, or $6 per ton. Peabody officials quoted "ballpark" figures of their extra charge for cleaned coal of $5 per ton for minimally cleaned coal, and $10 per ton for extensively cleaned coal.

Studies of Coal Cleaning Costs

Eight studies of cleaning costs were reviewed by INFORM; the conclusions of six of them follow.*

1. TVA/EPA (1979) calculated both capital and operating costs in pennies per pound of sulfur removed for three levels of cleaning:

	% Sulfur Reduction	Annual Capital Cost in ¢/lb Sulfur Removed	Annual O&M Cost ¢/lb Sulfur Removed
PCC I:	32	$.37	$.16
PCC II:	36	$.40	$.16
PCC III:	30	$.45	$.18

2. Gibbs and Hill (1977) found that moderate cleaning costs $2 per ton, or if the loss of Btus in the cleaning process is taken into account, $3.25 per ton. Minimal cleaning cost $.50, or with Btu loss, $.71 per ton. Extensive cleaning cost $5.70, or with Btu loss, $9.27 per ton.

The estimates for cleaning costs provided by the four other studies are compared as follows:

Study	Date	Average Cleaning Cost per Ton	Range of Cleaning Costs per Ton	Number and Range of Processes
3. Randy Cole/TVA :	1978	$3.47	$3.30 to $3.70	3/all moderate cleaning
4. Hoffman-Muntner:*	1978	$3.84	$3.35 to $4.83	8/minimal to extensive cleaning

*The other two studies were not included, as PENELEC estimated costs only to its own plants, and Hoffman-Muntner's 1976 figures are less up to date than its 1978 figures.

5. Buder & Clifford:	1979	$5.72	$3.88 to $7.63	7/minimal to extensive cleaning
6. PEDCo:	1980	$4.85	$4.40 to $8.41	8/minimal to extensive cleaning

*Unlike the other three studies, Hoffman-Muntner does not include an allowance for the Btus lost in cleaning. Whether or not to include the cost of Btus lost in cleaning is a point of contention; Hoffman-Muntner argues that doing so leads to an inaccurate appraisal of the costs of cleaning as the only reason companies clean their coal is to sell it. If it cannot be sold in its raw state, then it is worthless and worthless Btus lost cannot be calculated into the costs of cleaning coal.

Of the six studies reviewed, only Hoffman-Muntner (1978) bases its projections on actual preparation plants. PEDCo uses Hoffman-Muntner's data. Gibbs and Hill uses "representative industry experience," but its figures "do not reflect actual results in any particular situation." TVA/EPA, Randy Cole, and Buder and Clifford all base their projections on hypothetical preparation plants.

Hoffman-Muntner (and PEDCo) provide detailed case studies of the costs of eight physical coal cleaning plants. These plants represent a spectrum of technologies from a relatively simple jig process to complex circuits involving heavy-medium cyclones, froth flotation cells, and thermal dryers. The following tables provide a comparison of Hoffman-Muntner's and PEDCo's estimates.

BENEFITS OF BURNING CLEANED COAL

Of the eight studies consulted by INFORM that examined the benefits to utilities of burning cleaned coal, only PEDCo looked at the total range of benefits studied by INFORM. Using data from Hoffman-Muntner (1978), the only study to examine actual preparation plants, PEDCo found the cost of cleaning to be $4.85 per ton and the overall gain to utilities $7.20 per ton of cleaned coal; or if the cost of cleaning is subtracted, $2.35.* The other seven studies looked at sav-

*Buder and Clifford also look at overall benefits, but not at the full range considered by PEDCo and INFORM.

Hoffman-Muntner Case Study (1977)

Example	Process/Level	Input Capacity (tph)	Capital Cost Per Ton Per Hr. Input ($)	Clean Coal Output (tph)	Btu Recovery (%)	Operating and Maintenance Costs($)*		
						Per Ton Raw Coal	Per Ton Clean Coal	Per Million Btu†
1	Jig/Simple	600	6,600	354	91.6	1.97	3.35	0.138
2	Jig/Intermediate	1000	13,700	714	96.4	2.62	3.67	0.152
3	Jig/Intermediate	1000	12,100	566	83.0	2.22	3.92	0.157
4	Jig/Complex	1600	14,300	953	93.7	2.60	4.36	0.162
5	Heavy Media/Simple	1400	13,800	1,036	94.6	2.79	3.76	0.185
6	Heavy Media/Complex	600	22,400	440	89.2	3.54	4.83	0.177
7	Heavy Media/Complex	600	14,000	360	93.1	2.09	3.48	0.137
8	Heavy Media/Complex	900	23,200	774	94.3	2.91	3.38	0.135

*Includes capital amortization
†Does not include allowance for Btu loss of Process

PEDCo Case Study (1977)

Plant No.	O&M Cost,* $/ton of Cleaned Coal	Capital Charges† $/ton of Cleaned Coal	Cost of Btu Loss, $/ton of Cleaned Coal	Total Cost, $/ton of Cleaned Coal	$/100 Btu Recovered	$/ton of Ash Removed	$/ton of Sulfur Removed
1	2.70	0.65	2.14	5.49	0.227	9.92	1,746
2	2.55	1.12	0.75	4.42	0.183	13.39	271
3	2.67	1.25	4.49	8.41	0.338	15.71	344
4	2.96	1.40	1.60	5.96	0.222	10.47	789
5	3.20	0.56	1.10	4.86	0.239	27.69	244
6	3.04	1.79	2.21	7.04	0.258	30.53	187
7	2.12	1.36	1.76	5.24	0.206	9.55	1,000
8	2.44	0.94	1.02	4.40	0.176	52.28	421

*Operating and maintenance cost includes labor, supervision, overhead, supplies, fuel
†Includes allowance for Btus lost in cleaning

ings to one or more aspects of generating electric power with coal, including flue gas desulfurization systems, boilers, ancillary processes, and transportation. While the findings are often based on different assumptions and are difficult to compare, taken together they do suggest that coal cleaning can result in significant savings to power plants.*

While the studies conclude that burning cleaned coal has economic advantages, no reliable data has yet been obtained on the overall benefits to actual power plants that burn cleaned coal. None of the ten utilities interviewed by INFORM has quantified its savings from burning cleaned coal. Three of the eight studies INFORM consulted base their calculations on actual power plants, but they compare the costs of burning coals of different quality, not the costs of burning raw and cleaned coal.† One of these studies, EPRI (1980), examined plants that had burned both raw and cleaned coal, but the study did not successfully produce meaningful information.

There are several reasons why the exact amount a utility can save by burning cleaned coal is, at this point, very difficult to calculate. First, coal companies provided INFORM with almost no information on the extra charge for a ton of cleaned coal. Second, utilities do not generally demand cleaned coal; rather, they set specifications for the amount of sulfur, ash, moisture, Btu-content, and other constituents they will accept in a ton of coal. Third, if they have not previously burned raw coal from the same source, utilities burning cleaned coal have little basis for comparing costs and savings. Finally, the savings from burning cleaned coal can vary widely, depending on such site-specific factors as generating capacity. Plants with excess capacity will, of course, benefit less from increased boiler availability than will plants with a limited capacity.

---

*Bechtel Corporation recently published yet another cost study affirming the economic benefits of burning cleaned coal. In six out of seven two-unit 1,000 Mw power plants selected as study cases, the use of coal cleaning "reduced overall capital costs, and in five cases, the (operating) cost savings introduced by the use of coal cleaning more than offset the incremental cost of coal cleaning." In addition, Bechtel considers their findings to be "conservative" because not all of the possible economic benefits of using cleaned coal were included in their cost estimates.[36]

†These three studies are: Hoffman-Muntner (1978), EPA/TVA (1979), and EPRI (1980).

Flue Gas Desulfurization

Six studies examined by INFORM investigate the effect of using cleaned coal on FGD costs. Most of them suggest that using cleaned coal reduces the capital, and operating and maintenance costs of scrubbers, although the savings are difficult to define precisely.

Two of these studies found that power plants with older boilers (built before 1978) burning cleaned coal can, in certain cases, even eliminate the need for scrubbers.

1. The EPA/TVA (1979) study claims that, for coals with a sulfur content below 12 percent, cleaning is a "cost-effective" means of meeting a 1.2 lb $SO_2/10^6$ Btu standard. However, the study does not provide a comparison of the cost of using cleaned coal with the cost of using scrubbers.

2. PENELEC conducted an in-house study in 1977 which concluded that the use of cleaned coal versus FGD (from its Homer City preparation plant, which produces two grades of coal: intensively and moderately cleaned), could save the utility $28,340,000 in capital costs and $10,405,000 in operation and maintenance costs on its three 600 Mw boilers (see PENELEC profile). However, the costs of the preparation plant which PENELEC has since installed and retrofitted have risen significantly beyond those anticipated in the study.

Five studies examined by INFORM compared the costs of using cleaned coal in combination with scrubbers with the cost of using scrubbers alone. They noted potentially large savings to FGD capital, and operation and maintenance costs. As two of the studies [EPA/TVA (1979) and PEDCo (1980)] indicate, these savings are not always large enough in themselves to offset the cost of cleaning. However, if the benefits to the other components of power generation are taken into account, using cleaned coal with scrubbers is *generally* less expensive than using scrubbers alone.

The conclusions of these studies are difficult to compare directly, largely because they are based on different assumptions of emission standards, the sulfur content of the coal being burned, and the extent to which the coal has been cleaned. Although some of the studies tried to anticipate the Revised New Source Performance Standards, all used different projections to do so; projections that do not reflect the actual standards that were issued.

1. Hoffman-Muntner (1976) compared the cost of using a scrubber alone with the cost of using a scrubber in combination with precleaned coal and found considerable savings in both FGD capital, and operating and maintenance costs if cleaned coal is used. For a new 500 Mw power plant, meeting a 1.2 lb $SO_2/10^6$ Btu standard would be from 16 to 50 percent less expensive if cleaned coal is used with scrubbers; meeting a .55 lb $SO_2/10^6$ Btu would cost these plants 48 percent less. For existing plants, using scrubbers alone to meet this standard would be from 13 to 140 percent more expensive than with coal cleaning.

2. Randy M. Cole/TVA (1979) compared the costs of 1,000 Mw power plants burning three types of coal under an emission standard of 85 percent sulfur dioxide reduction. He concluded that reduced FGD capital costs more than offset the capital and operating costs of coal preparation.

3. Buder and Clifford (1979) studied the economics of combining scrubbers and precleaned coal, using seven types of coal and three levels of coal cleaning for 1,000 Mw power plants with twin 500 Mw units. The study assumed an emission standard of 85 percent sulfur dioxide removal, with a 1.2 lb $SO_2/10^6$ Btu ceiling and a .2 lb $SO_2/10^6$ Btu floor. The capital costs of scrubbers were found to be lower when cleaned coal is burned.

   In fact, as the table below indicates, the more intensive the cleaning, the lower the FGD capital costs:

Level of Cleaning	FGD Capital Cost*
2A (least intensive)	129.3
2B	124.7
2C	121.8
7A	123.6
7B	119.4
7C (most intensive)	117.1

*in millions of January 1979 dollars.

Buder and Clifford also suggested that using cleaned coal could result in considerable savings to FGD operation and maintenance costs, but they did not quantify these savings.

4. EPA/TVA (1979) found that the less stringent the sulfur dioxide emission standard, the more cleaned coal can benefit FGD costs. Conversely, "the economic advantages of coal cleaning alone or with FGD, decrease as the allowable emission level is lowered," ultimately to the point where it is no longer an advantage. The study found that using cleaned coal in combination with scrubbers costs more than using scrubbers alone when a coal with a sulfur content of more than 3 percent had to meet an 85 percent sulfur dioxide reduction requirement. However, for coal with a sulfur content of less than 3 percent, it was less expensive to meet this standard by using FGD and cleaned coal together.

The study also noted that when non-FGD operating costs were taken into account, using cleaned coal with FGD was less expensive than using FGD alone. Furthermore, FGD systems "have experienced markedly better operation, in addition to a reduction in maintenance requirements" when low rather than high sulfur coal is used. However, the TVA/EPA study did not quantify these benefits.

5. PEDCo (1980) found that when cleaning removed 32 pounds of sulfur per ton of raw coal (approximately a 30 percent reduction), the total benefit to utilities was $7.20 per ton of cleaned coal. Of this amount, approximately $4.00 per ton was savings derived from the capital, and operation and maintenance of scrubbers. However, like EPA/TVA, PEDCo found that very restrictive emission standards could reduce these benefits to almost nothing, while less restrictive standards could, in some cases, make the benefits of FGD large enough to offset the cost of cleaning.

Boilers

Although the data from INFORM's sample and from the studies we consulted clearly indicate that the use of cleaned coal results in reductions in boiler maintenance costs, quantitative data is limited. As a 1980 EPRI study found, utilities generally do not consider the benefits to boilers when trying to assess the savings resulting from burning cleaned coal. In spite of the limited statistics available, the studies INFORM consulted found a relationship between coal quality and boiler performance. One can infer that cleaning coal, which removes ash and sulfur and generally improves coal quality, can also provide sufficiently large benefits in boiler operation to be a major factor in off-setting the costs of using cleaned coal.

1. Holmes (1969)* was the first to investigate the impact of coal quality on the operation and maintenance of large central station boilers. He examined two TVA plants with virtually identical equipment that burned coals of different quality, and discovered that boiler maintenance costs per ton of coal were nearly twice as high for the plant burning the inferior coal.

2. Using the TVA data from the Holmes study, Hoffman-Muntner (1976) showed that the cost of maintaining a boiler could be related to the amount of sulfur and ash in the coal:

THE EFFECT OF ASH AND SULFUR CONTENT
ON BOILER MAINTENANCE COSTS

Additive Reduction in Ash and Sulfur, (%)	Maintenance Cost Savings, $/ton Coal
15	0.33
12 - 15	0.30
9 - 12	0.27
7 - 9	0.24
5 - 7	0.20
3 - 5	0.17
2 - 3	0.13

*Not one of the ten studies included in INFORM's survey.

Using the relationship postulated between maintenance costs and coal quality, Hoffman-Muntner estimated the savings from burning cleaned coals of varying quality in 50 hypothetical and real plants. The study found savings in boiler maintenance costs to range from $1.97 to $2.16 per ton of cleaned coal.

3. Cole (1978) and Phillips (1979) determined that burning coals with a higher sulfur and ash content resulted in added boiler operation and maintenance costs and in additional costs associated with a loss of peaking capacity and reduction in boiler availability. As the table below indicates, such costs increased rapidly from $.0/ton for a combined sulfur and ash content of 12.5 percent, to $6.41/ton for a combined sulfur and ash content of 25 percent.

COST PENALTIES ASSOCIATED WITH ASH AND SULFUR
CONTENT OF THE COAL

Ash Content (%)	Sulfur Content (%)	Total A&S (%)	Cost Penalty, $/ton Coal Fired				
			Maintenance Costs	Peaking Capacity	Rated Capacity	Plant Avail.	Total
10.5	2.0	12.5	0	0	0	0	0
12.5	2.5	15.0	0.38	0	0	0	0.38
14.5	3.0	17.5	0.75	0	0	0	0.75
16.5	3.5	20.0	1.13	0.19	1.08	0.47	2.87
18.5	4.0	22.5	1.50	0.23	2.08	0.91	4.72
20.5	4.5	25.0	1.88	0.21	3.00	1.32	6.41

4. PEDCo (1980) concluded that cleaned coal's benefits to boiler performance were second only to its benefits to scrubbers. PEDCo found that, along with its reduced ash and sulfur content, cleaned coal's reduced mineral content (iron, sodium and calcium) decreased boiler fouling and slagging. As the table below indicates, the resulting improvement in boiler availability and capacity can provide large savings.

SAVINGS IN BOILER OPERATION
(Dollars per ton of cleaned coal)

Benefit area	Typical	Range
Boiler Operation		
Operating & Maintenance	$0.40	$0.10 - $2.00
Availability	$1.90	$0.30 - $5.10
Efficiency	$0.10	$0.25 - $0.50
Capacity	$0.00	$0.00 - $9.00
TOTAL	$2.40	

5. Hoffman-Muntner (1980) was the first study to look at the separate effects of ash and sulfur content on boiler maintenance costs. Earlier studies had looked only at the combined effects. Hoffman-Muntner examined the records of maintenance costs for five TVA boilers over an extended period from 1962 to 1978. The study found that each percent of ash removed reduced the cost of burning a ton of coal by $0.05 to $.10 in 1978 dollars. Each percent of sulfur removed reduced the cost by $.30 to $.50 per ton in 1978 dollars.

6. EPRI (1980) was the only study consulted by INFORM that looked at power plants that have burned raw and then cleaned coal at the same plant. Other studies have not done this, largely because very few utilities have operating experience with using the same coal both raw and cleaned.

EPRI studied Commonwealth Edison's Kincaid plant and Pennsylvania Power and Light's Montour station. However, it did not get usable information from either plant. The Kincaid data could not be used because, before the switch from raw to cleaned coal, the boiler had deteriorated significantly and major modifications had to be made. At the Montour plant, coal quality was found to have no statistically discernible effect on boiler efficiency. Other factors, such as heat rate, internal power consumption, and availability, could also not be correlated with coal quality. EPRI concluded that the true relationship between coal quality and boiler performance could not be determined because other variables exerting a greater influence masked the improvements resulting from better quality coal. The study also concluded that neither the current data nor the current analytical tools are sufficiently developed to allow for reliable quantification of coal cleaning's benefits.

Ancillary Processes

The use of cleaned coal has been shown to have both positive and negative effects on the costs of coal handling and storage pulverization, ash handling and storage and particulate collection.

Coal Handling and Storage

PEDCo (1980) attempted to quantify the cost savings in handling and storage resulting from using cleaned coal. These savings were found to be negligible: only about $.01 per ton of cleaned coal.

Pulverization

Cleaning reduces the amount of coal to be pulverized and, by removing some of the harder impurities such as sulfur, ash and rock, makes coal easier to grind. The studies INFORM reviewed found that in general, using cleaned coal reduces capital, and operation and maintenance costs for pulverizers and increases pulverizer capacity.

1. In his study of four TVA power plants (1969), Holmes found a significant correlation between the cost of maintaining pulverizers and the ease with which the coal feed can be ground. Pulverizer maintenance costs increased linearly as the coal's sulfur and ash content increased. When the combined sulfur and ash reached 17.5 percent, pulverizer maintenance costs increased at a higher rate.

2. In case studies of twelve new and existing power plants using coals of different quality, Hoffman-Muntner (1976) found pulverizer savings to be between $.01 and $.04 per ton of cleaned coal, with an average saving of $.03.

3. Gibbs and Hill (1978) noted similar savings, estimating that pulverizer maintenance, normally budgeted at $.05 to $.06 per ton, could be cut in half if cleaned coal were used. They also estimated that the reduced tonnage of cleaned coal required to feed the boiler would result in savings of $.50 per ton of coal pulverized.

4. PEDCo (1980) estimated that increased pulverized capacity from reduced coal tonnage would save from $0.0 to $4.50 per ton of cleaned coal. Plants that already have adequate pulverizer capacity would not benefit, while those with limited capacity could realize large benefits. (PEDCo based its estimates on the capital that would be required to generate additional megawatts of electricity now provided through the use of cleaned coal.)

Ash Handling and Storage

PEDCo (1980) found that coal cleaning reduces ash handling and maintenance costs slightly: an average of $.05 per ton of cleaned coal, with an upper limit of $.10 per ton. The study assumed that ash handling accounted for 16 percent of all operation and maintenance costs, and that ash handling requirements are reduced in direct proportion to reductions in the coal's ash content. PEDCo also examined the costs of ash and sludge disposal at eight utilities. Assuming a 50 percent reduction in ash content, PEDCo estimated benefits ranging from $.00 to $.25 per ton of cleaned coal, with an average of $.10 per ton. The study also noted, however, that since cleaning reduces more ash than sulfur, a net deficit of fly ash could result, making it more difficult to stabilize the sludge produced in scrubbing.

Particulate Collection

Cleaned coal has a mixed impact on the cost of particulate collection.

1. PEDCo (1980) found that the required electrostatic precipitator (ESP) size may be lowered by 5 to 25 percent if cleaned coal is to be burned. PEDCo found that the capital savings apply only to new plants where size reductions are made at

the design stage. In a new 1,000 Mw plant, the cost of an ESP, which accounts for 5 percent of the total cost of the plant, could be reduced by 10 percent. These benefits were calculated to be an average of $.45 per ton of cleaned coal, with a range of $.25 to $1.00 per ton. PEDCo noted that these estimates agreed with estimates of $.84 per ton savings made by Buder (1979).

2. Gibbs and Hill (1978) found that, for a new 500 Mw plant which must meet an emission standard of .10 lb $SO_2/10^6$ Btu, using cleaned coal could result in capital savings. Cleaning high-sulfur coal could save an average of $.11 to $.13 per ton for level C and D cleaning, respectively. Cleaning low-sulfur coal could save $.20 or $.23 for level C and D, respectively.

Transportation

Three studies reviewed by INFORM have attempted to determine the savings in transportation expenses of utilities buying cleaned coal. All of them found that the reduced tonnage involved in shipping cleaned coal saved utilities money, but the amount varied considerably.

1. Hoffman-Muntner (1976) studied twelve mine and power plant pairs that are an average of 436 miles apart. They found that transporting cleaned coal saves between $.15 and $.45, or an average of $.20 per ton.

2. Buder and Clifford (1979) estimated that transporting cleaned coal could save utilities from $.001 to $0.18 per net Kwh, depending on the distance shipped and the increase in Btus per ton of coal.

3. PEDCo (1980) estimated that transportation savings would range from $.10 to $1.50 per ton of cleaned coal and average $.70 per ton.

Part III

Coal Cleaning Technologies

Chapter 4

Methods of Coal Preparation

Coal preparation technologies have been used for over 100 years. Since the passage of the Clean Air Act in 1970, however, they have taken on greater significance as a limited means of removing sulfur from coal prior to combustion and reducing emissions of polluting sulfur dioxide. In general, coal preparation yields a more uniform quality fuel of greater heating value (Btu's) by reducing ash and trace elements (in some instances), removing mining wastes, and concentrating fixed carbon.

"Coal preparation" is frequently used loosely to describe all of the seven processes that coal undergoes between mining and combustion. These include transportation, storage, crushing, classification by size, separation of impurities, removal of water, and disposal of waste. However, for the purposes of this study, the definition of "coal preparation," also commonly termed "benefication," "cleaning" or "washing,"* will be narrowed to only those physical and chemical processes that upgrade the quality of the fuel by regulating its size and reducing the quantities of ash, sulfur and other impurities found in it.[1]

---

*In the past, "cleaning" referred to the dry coal preparation technologies and "washing" referred to the wet coal preparation technologies. Today, the two terms are used interchangeably.

The chemical coal cleaning technologies promise greater total sulfur removal, although they are neither technically nor economically feasible at this time (for further discussion of the chemical coal cleaning process, see the Chemical Process chapter). Dry physical coal cleaning technologies, such as pneumatic cleaning and electrostatic separation, have also been used in the past without great commercial success.[2] Further, few of the dry cleaning technologies have demonstrated commercial potential, with the possible exception of the magnetic separation technologies (for further discussion of dry physical coal cleaning and magnetic separation technologies, see the Advanced Physical Coal Cleaning chapter).

At the present time, only the physical coal cleaning processes are used commercially.

Physical Coal Cleaning Processes

Today, coal preparation plants incorporate a variety of systems, ranging from those designed simply to remove coarse refuse from raw coal to sophisticated systems designed to remove the maximum amount of pyritic sulfur and ash. Thus a coal cleaning plant does not employ a specific process, but rather a number of different operations applied sequentially or in various combinations. A modern coal cleaning plant is a "continuum of technologies rather than one distinct technology."[3]

Coal to be cleaned is usually separated into three size fractions: coarse, intermediate and fine.[4] Preparing coals for more extensive cleaning involves crushing the raw coal feed to smaller size fractions and cleaning the coal particles of smaller and smaller size fractions.[5] Additionally, each fraction can be treated with more specific and sophisticated methods and equipment.

The six major operations are crushing, sizing, cleaning, dewatering, clarifying and drying.[6,7] The equipment available to accomplish these operations is produced by different manufacturers, and each unit has its own site specific applications. These different operations and variations are described below.

Crushing

Crushing, or comminution, liberates the larger impurities embedded in raw coal, and provides a product of more uniform size, facilitating the ease of transportation and proper combustion.[8] Over the years, there has been a great increase in the amount of coal which is crushed after mining.[9] In 1940, only 8 percent of the total U.S. bituminous coal production was mechanically crushed.[10] By 1975, this figure had risen to 90 percent.[11] Recently there has also been a general increase in the size of crushing and grinding equipment, which has made it possible for larger volumes of coal to be cleaned.

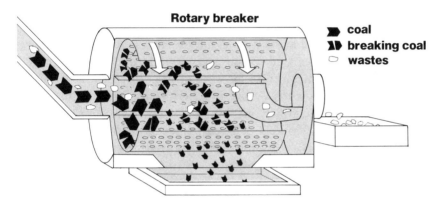

Sizing

Sizing separates feed coal into various fractions suited for washing by different cleaning processes. Sizing can be accomplished by wet or dry screening or classification equipment, which use variations in the flow rates of different size particles through a fluid or air. Sizing of particles above 28 mesh ($\frac{1}{2}$ mm) is generally performed on vibrating screens. For smaller particles of coal, sieve bends and classifying cyclones are used.[12,13,14]

Cleaning

The specific gravity of clean coal is less than that of ash, pyritic sulfur or the other impurities found in raw coal.* This difference in

---

*Most coals have a specific gravity that ranges from 1.12 to 1.70. Ash and other impurities have a wide range of specific gravities: Pyritic sulfur—4.6 to 5.2; gypsum, kaolinite and calcite (typical ash constituents)—2.3, 2.6, and 2.7; and sandstone, clay and shale (typical mining wastes)—2.6.[15,16]

specific gravity is used by almost all of the commercially-used wet physical coal cleaning technologies to separate larger-sized coal particles (larger than ½ mm) from their associated impurities.

These specific gravity-based technologies, however, are usually not effective in cleaning raw fine-size coal particles smaller than ½ mm in size. When the size of coal is reduced, the surface area of the resulting fine coal particles is increased. Because of this increase in surface area, the separation times are longer for the fine particles. The pyrite in coal is often found in small, highly disseminated particles requiring extensive size reductions of the raw coal to facilitate its removal.

Seven processes are commonly used to clean coal:

Jigs are one of the oldest, and still the most common technology used for washing, separating and concentrating mineral ores. Introduced in Pennsylvania in 1873 for washing coal, jigs are used today to treat materials with a wide range of specific gravities, including gold (which has a specific gravity of 19.0). Jigging involves stratifying raw coal in water by pulsations that rearrange the particles so that the lightest (the coal) move to the top of the bed, while the heaviest particles (the impurities) settle to the bottom; the two products can then be collected separately.[17]

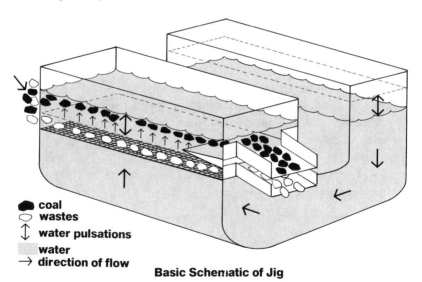

● coal
○ wastes
↕ water pulsations
▨ water
→ direction of flow

Basic Schematic of Jig

Innovations in design have kept jigs competitive with other cleaning devices. For example, the Batac jig, which was recently introduced in the United States, can process finer-sized coals more efficiently than the conventional Baum jig.[18]

Dense-media vessels use liquids that are heavier than water to separate coal from its impurities. Particles of coal as small as ¼ inch in size can be treated with this process. Finely ground magnetite (or other materials) mixed with water is used to make a slurry with the specific gravity required for the particular separation between coal and refuse. The specific gravity of the separating medium can be adjusted by varying the concentration of the magnetite to compensate for the variability in the specific gravities of the impurities contained within the coal, permitting the separation at the specific gravity needed to achieve the desired result.[19] The specific gravity can be varied to maintain a constant product, even with variation in the amount of impurities contained in the raw coal.

Principle of Dense Media Vessel*

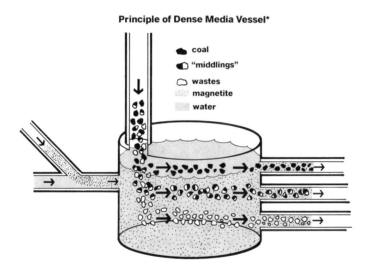

coal
"middlings"
wastes
magnetite
water

*artist's rendering from McNally-
Pittsburg publication, "Btu," May 1972, p. 10.*

The use of this process is second only to the use of the jig. However, it is gaining favor because very efficient separations between coal and refuse can be obtained by its use. In 1948, only one commercially-operated coal plant in the U.S. used a heavy-medium process. By 1963, at least 115 plants in the United States were using this equipment to treat 49 million tons of coal annually.[20]

Magnetite recovery is an important part of this process, as nearly a pound of magnetite is used for each ton of raw coal. The amount of magnetite used depends largely on the size of the coal cleaned; the smaller the size of the particles to be cleaned, the greater the amount of magnetite consumed. The 1981 market value for magnetite was approximately 5¢ per pound.[21]

Dense-medium cyclones (revolving receptacles) are used to effect a sharper separation of intermediate sized raw coal material ($\frac{1}{2}$ inch x $\frac{1}{2}$ mm). In this process, the gravity separation of a dense-media mixture (described above) is enhanced by centrifugal force.[22,23]

Dense Media Cyclone*

coal
wastes
centrifugal action
magnetite
water

*artist's rendering from McNally-
Pittsburg publication, "Btu," May p. 12.

Hydrocyclones or water-only cyclones also enhance separation by centrifugal force, but they do not use a dense-media mixture. They are now being used more often in new preparation plants as rougher separating devices that decrease the load on downstream cleaning equipment. They are also used to clean very small flotation size coal (-28 mesh).[24,25] Hydrocyclone equipment is advantageous because it can process large tonnages of raw coal at a relatively low cost, and it requires comparatively less space and maintenance.[26] However, once installed, its power and water requirements are high.[27]

The concentrating table, a common hydraulic separator, is one of the oldest and most widely used cleaning devices in the United States. It is used especially for processing particles in the ⅜ inch to 0 size range. This method takes advantage of the fact that feed coal particles, "shaken" across a nearly horizontal surface, will separate from refuse because of their different densities.[28]

The froth flotation process is used to clean fine coal particles of less than ½ mm in size. Technologies used to remove the impurities associated with fine coal particles take advantage of the fact that coal, like oil, has surface properties that make it "hydrophobic" or water-hating. The most important of these technologies is froth flotation. In this process, raw fine coal particles are immersed in a water bath, through which air is bubbled. The water-hating coal particles cling to these air bubbles (nearly always with the help of flotation reagents) and can be skimmed off in a surface froth. The heavier coal impurities, which are not so hydrophobic, sink and mix with the water.[29]

Principle of Froth Flotation*

froth sticking to coal
coal
wastes
water

artist's rendering from McNally-Pittsburg publication, "Btu," May, 1972, p. 14.

The oil agglomeration process, which is still in the research and development stage, utilizes the same water-hating properties of coal particles, but uses oil droplets instead of flotation reagents. (This technology will be discussed in greater detail in the Ames Laboratory/Iowa State University profile in the advanced physical coal cleaning section.

Pneumatic cleaning processes such as air tables currently have limited application. Use of pneumatic cleaning equipment and methods decreased from about 14 percent in 1940 to about 2 percent in 1977 as a result of problems associated with dust, plant emissions, the higher moisture content of today's raw coals and the size limitations of pneumatic cleaning equipment.[30] But these methods of coal cleaning are attracting new interest in the west, where water is short and the coal contains fewer impurities. However, a large amount of research and development is needed to improve the efficiencies of these processes, and commercialization is not economically feasible at this time.

Dewatering

Wet physical coal cleaning processes leave a significant amount of moisture on the coal. This additional moisture decreases the fuel value of the coal, increases transportation and handling charges by adding to the weight of the coal shipment, and can make the coal prone to freezing in winter, causing problems in handling and use. To reduce the water content of the coal, preparation plants (in many cases) first dewater the coal mechanically, and then dry it by direct heating.

Screens are the most common type of fine coal dewatering equipment, and improved design in recent years has enhanced their dewatering capabilities. The "sieve bend" is now used throughout the world for preliminary dewatering before the coal passes to the vibrating screens and centrifuges. Another screen gaining popularity as a dewatering device is the "Vor-Siv," developed by Polish Coal, Inc. It is a stationary device, capable of handling high volumes of solids in a water slurry.[31]

Centrifugal dryers are used to dewater coal particles 1½ inch and smaller by spinning in a centrifuge. The effectiveness of this equipment depends upon the centrifugal force developed, the size of the particles, and the length of time the coal is left in the dryer.[32,33,34]

The vacuum filter is the most common dewatering device for fine-sized coal particles. Ninety percent of these filters are of the continuous vacuum type, which work well, are less expensive to purchase and install per square foot of filtration area and require less space than other filters. Solids form a cake on the leaves of this filter and are removed by blasts of air.[35]

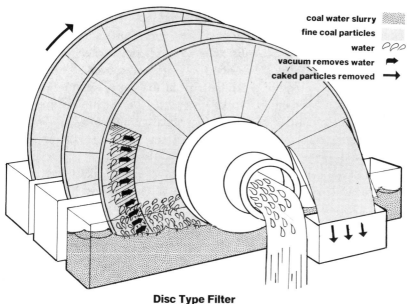

coal water slurry
fine coal particles
water
vacuum removes water
caked particles removed

Disc Type Filter

Clarifying

Water Clarifiers are used in most preparation plants to remove nearly all of the solid impurities from process water. Preparation plants have very high recirculating requirements of up to 16,000 gallons per minute for a plant cleaning 1,200 tons of coal per hour (e.g. the Homer City plant). The total water requirements for all U.S. preparation plants is hundreds of billions of gallons per year.[36] Eighty percent of this water is recycled, and overall water management is of key importance.[37]

Currently, all wet coal preparation plants have some type of partially closed water circuit. Recirculation and clarification of wash waters is a critical part of cleaning plant operation because such treated waters decrease the loss of fine coal in the waste, prevent the build up of solids in the water recirculated to the cleaning units, and help the plant comply with state and local laws regarding stream pollution—which have become increasingly stringent in the last several years (see Appendix: Coal Preparation Plant Waste Disposal for a discussion of regulations governing preparation plant effluent streams). In the past, some preparation plants bled off a large portion of the polluting water into streams to keep the level of solids within the system low.

The gravitational clarifier (or thickener/clarifier) is the most commonly used piece of equipment for water clarification. It is a large circular basin in which solids settle to the bottom where they can be pumped away, leaving the clarified water to be removed at the top. The nature of the waste slimes and the need for clear water require that chemicals for flocculation be added to the water in the clarifier. This results in the formation of "flocs" (or clumps of wastes) and an increase in the settling rates of solids. Unless large amounts of clay are present in the wastes, requiring greater quantities of flocculates, the costs of these chemicals are generally more than adequately paid back by the reduced size of clarifiers and the enhanced clarity of plant water.[38,39]

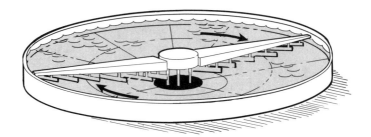

Static thickener

Cyclones are used for clarifying plant water by controlling the solids concentration of the circulating water. A mixture of fine coal and water is concentrated by the centrifugal force of the cyclone to produce a thick suspension, which undergoes additional dewatering in filters or centrifuges.[40]

Thermal Drying

Thermal drying equipment removes moisture by passing hot gases over coal. It can absorb the surface moisture from cleaned coal, although the internal moisture is largely unaffected. Prior to 1967, the use of thermal dryers was increasing.[41] Since then, however, the number of dryers employed in preparation plants decreased as a result of stricter dust emission regulations which significantly increased the cost of the process. Only 11.8 percent of the mechanically-cleaned coal in the United States was thermally dried in 1977.[42]

Seven different types of thermal dryers are currently in use: the fluidized-bed,* multilouver, rotary dryer, screen dryer, suspension or flash dryer, and the vertical tray and cascade dryer.[43] The fluidized-bed dryer, introduced to the coal industry in 1955, represented more than half of the drying units in operation in 1981, processing over 72 percent of the coal to be dried.[44]

Levels of Coal Preparation

Not all coals are good candidates for coal cleaning, owing to differences in their constituents and physical properties. Also all coal preparation technologies have several problems and limitations that may reduce potential applications.

Depending upon the physical composition of the raw coal and the market requirements for cleaned coal, individual preparation plants may employ different combinations of the methods and equipment described above. The more fine coal particles cleaned, and the more sulfur, ash and other impurities removed, the more rigorous and expensive the levels of coal cleaning.[45]

The extent to which coal is cleaned divides all the coal preparation processes into different levels. There are many different classification schemes defining these possible levels. This report divides coal cleaning technologies, but they are actually only points along a continuum of possible levels of coal cleaning. In the real world of coal preparation, the distinctions between "levels" of coal preparation are often blurred.

The following table summarizes the five levels of coal preparation identified by INFORM.†

---

*See INFORM's study, *Fluidized-Bed Energy Technology: Coming to a Boil*

†The five levels described in this section are based on categories developed by Gibbs and Hill, Inc. in *Coal Preparation for Combustion and Conversion* (Palo Alto, California: Electric Power Research Institute, May 1978), pp. 2-49 through 2-64.

Greenwich Collieries Preparation Plant (simplified flow design)

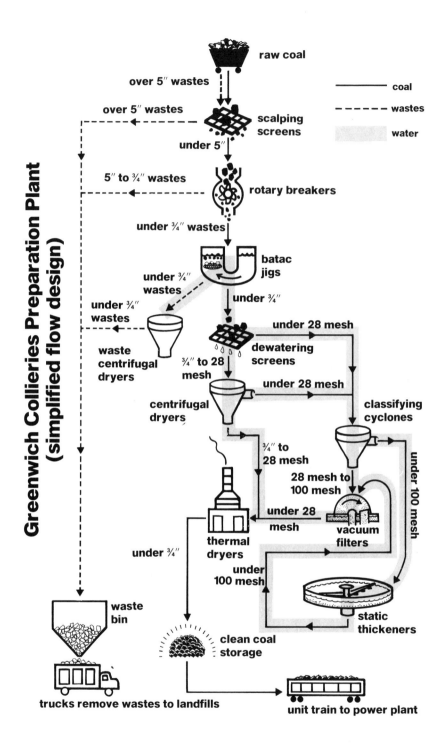

raw coal

coal
wastes
water

over 5″ wastes

over 5″ wastes

scalping screens

under 5″

5″ to ¾″ wastes

rotary breakers

under ¾″ wastes

batac jigs

under ¾″ wastes

under ¾″ wastes

under ¾″

under 28 mesh

waste centrifugal dryers

dewatering screens

¾″ to 28 mesh

under 28 mesh

classifying cyclones

centrifugal dryers

¾″ to 28 mesh

28 mesh to 100 mesh

under 100 mesh

under 28 mesh

thermal dryers

vacuum filters

under ¾″

under 100 mesh

waste bin

static thickeners

under ¾″

clean coal storage

trucks remove wastes to landfills

unit train to power plant

Level 1 — Crushing and Breaking

Process Description:

Coal crushed to less than three inches (-3 ") for top size control.

Typical Percent of Raw Coal Weight Recovered:

98% to 100%

Percent of Raw Coal Btu Content Recovered:

99% to 100%

Typical Reduction Potential:

Ash: None to minor

Sulfur: Almost none

Benefits:

Removes very large-sized impurities such as wood, large rocks.

Smaller-size coal is better suited to transportation and handling needs.

Btu content of cleaned coal is almost 100% that of raw coal†.

Liabilities:

Ash and sulfur removal almost nonexistent.

Little increase in the heating value of the cleaned coal.

---

†When coal is cleaned, some of the Btu content or heating value of the raw coal is lost in the refuse that is removed from the cleaned coal. However, the Btu value of the raw coal per pound is usually *less* than the Btu value of the cleaned coal per pound; this is due to the concentration of combustible materials in clean coal by the removal of ash and rocks that do not contribute to the heating value of the fuel. Thus, the Btu value per pound of cleaned coal may be higher than that of raw coal, although a portion of the *total* Btu content of the raw coal is lost in the discarded waste materials.

Level 2 — Coarse Coal Cleaning

Process Description:

Coal crushed to -3″.

3″ x ⅜″ "coarse" coal cleaned in equipment based on principles of specific gravity.

-⅜″ coal particles removed from raw coal and re-combined with 3″ x ⅜″* cleaned coal prior to shipping.

Percent of Raw Coal Weight Recovered:

75% to 85%

Percent of Raw Coal Btu Content Recovered:

90% to 95%

Typical Reduction Potential

Ash: Fair to good

Sulfur: None to minor

Benefits:

3″ x ⅜″ mining wastes (rocks) and some ash removed.

Heating value of the cleaned coal raised slightly.

Liabilities:

Little, if any, inorganic sulfur removed; coal not cleaned in small enough size-fractions to enable release of pyritic sulfur, except for large-sized pyritic sulfur impurities.

-⅜″ coal particles remain uncleaned and shipped with final cleaned coal product.

---

*The sizes of coal particles (size-fractions) given in these tables are representative of those commonly found within the industry where size-fractions vary in different plants. Here, the term 3″ x 3/8″ refers to the coal that is between 3 inches and 3/8 inch in size. This abbreviation is used throughout these tables.

Level 3 — Moderate Coal Cleaning

Process Description

Coal crushed to -3".

Separated into three size fractions: 3" x ¼", ¼" x 28 mesh* and -28 mesh.

3" x ¼" coal cleaned in equipment based on principles of specific gravity.

¼" x 28 mesh particles receive more elaborate cleaning in specific-gravity equipment.

-28 mesh particles can be added uncleaned to other, cleaned fractions or discarded if they contain high percentages of sulfur and ash.

Percent of Raw Coal Weight Recovered

60% to 80%

Percent of Raw Coal Btu Content Recovered

80% to 90% †

Typical Reduction Potential

Ash: Good

Sulfur: Fair

---

*Mesh is a measure of size often used to describe small coal particles. A 28 mesh sieve has 28 holes or wires per linear inch, each opening equivalent to .595 mm. or .0234 inch. "-28 mesh" refers to all coal particles that are smaller than 28 mesh. When smaller sized coal particles with increased surface areas are washed, the moisture content of the cleaned coal increases geometrically. This increase in moisture may require thermal drying, which is energy intensive and therefore costly.
†More of the raw coal Btu content can be recovered if the -28 mesh fine particles are added to the final product.

Benefits

Cleaning of small size fractions accomplishes some inorganic sulfur removal.

Significant amount of ash and mining wastes removed.

Heating value of the cleaned coal raised significantly.

Liabilities

If -28 mesh uncleaned particles are added to cleaned coal, some sulfur and ash is reintroduced to the final product.

If -28 mesh uncleaned coal particles are not added to the final product, the Btu losses are high.

Level 4 — Fine Coal Cleaning

Process Description

Same as Level 3, except that -28 mesh particles are cleaned instead or discarded or added uncleaned to other cleaned fractions.*

-28 mesh particles cleaned either by specific gravity or surface property-based processes.

Percent of Raw Coal Weight Recovered

60% to 80%

Percent of Raw Coal Btu Content Recovered

80% to 90%

Typical Reduction Potential

Ash: Good to Excellent

Sulfur: Fair to Good

---

*In some Level 4 plants, raw coal is crushed to -3/4 " (instead of -3 ") in order to release greater quantities of ash and organic sulfur from the raw coal.

Hamilton Preparation Plant—Fine Coal Circuit
(simplified flow diagram)

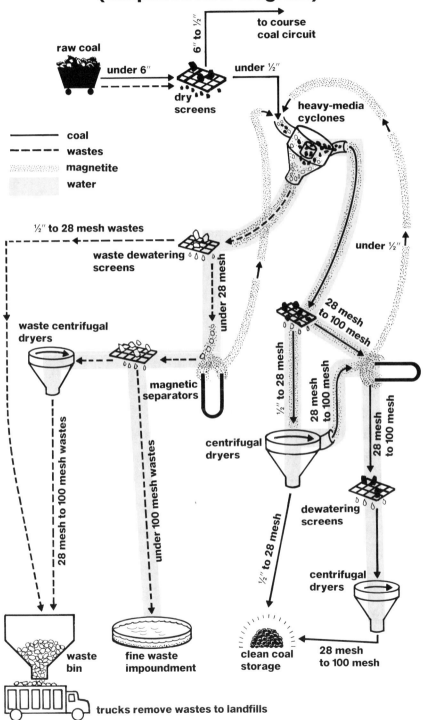

Benefits

Significant percentage of inorganic sulfur removed.

Most of the ash (i.e., extraneous ash) is removed.

Heating value of the cleaned coal significantly raised.

Liabilities

Cleaning of smaller coal sizes results in increased costs (equipment, and operating and maintenance costs) and increased quantities of wastes that require disposal.

Moisture-related problems, as with Level 3; thermal drying may be required.

Btu's are lost in waste materials.

Level 5 — Multiple Stream Coal Cleaning

Process Description

Raw coal crushed and cleaned in 3″ x ¼″, ¼″ x 28 mesh, and 28 mesh x 0.

Two distinct grades of cleaned coal produced: "deep cleaned," which is extremely low in sulfur and ash coal, and "middlings," higher sulfur and ash coal.

Low specific gravities used in "deep-cleaning" equipment to clean fine and ultrafine coal particles.

Percent of Raw Coal Weight Recovered

60% to 80%

Percent of Raw Coal Btu Content Recovered

85% to 95%

Typical Reduction Potential

Ash: Deep-cleaned coal—excellent; middlings coal—none to fair.

Sulfur: Deep-cleaned coal—excellent; middlings coal—none to fair.

Benefits

In some instances, deep-cleaned coal can be burned in plants that must meet stringent sulfur dioxide emission standards; middlings coal can be burned in plants that do not require extremely low sulfur or ash content in the fuel. The inorganic sulfur and ash content of most deep-cleaned coal is very low.

Fewer Btu's lost in wastes with Level 5 cleaning, as compared with levels 3 and 4; fewer combustible materials end up in wastes because the middlings, a higher sulfur and ash coal, is also recovered (an important consideration with costly coal).

Liabilities

High capital and operating and maintenance costs, relative to other levels of coal cleaning.

Problems related to use of a new technology.

Fine-coal related problems, as with Level 4.

Table A - LEVELS OF COAL PREPARATION

Level	Raw Coal Weight(%)	Raw Coal Btu Content(%)	Reduction Potential Ash	Sulfur	Comments
1	98 to 100	99 to 100	None to minor	None	Crushing and breaking of raw coal to 3" size
2	75 to 85	90 to 95	Fair to good	None to minor	Coarse coal cleaning of 3" X 3/8" coal
3	60 to 80	80 to 90	Good	Fair	Moderate coal cleaning of 3" X 28 mesh coal
4	60 to 80	80 to 90	Good to excellent	Fair to good	Fine coal cleaning of 3" X 0 coal
5	60 to 80	85 to 95	Deep-cleaned coal: excellent; middle-cleaned coal: none to fair		Multiple-stream coal preparation of two cleaned coal products: "deep-cleaned" coal and "middle-grade" coal

Chapter 5

Advanced Physical Coal Cleaning

Advanced physical coal cleaning technologies are particularly innovative variations of the ''conventional'' physical coal cleaning processes that remove impurities from raw coal by crushing the fuel and treating it in a water slurry with jigs, cyclones, froth-flotation cells, and other equipment.

Advanced physical coal cleaning promises several advantages over conventional technologies: (1) higher percentages of ash and inorganic (or pyritic) sulfur can be removed; (2) hard-to-clean fine coal particles can be treated more effectively; (3) costly and energy-intensive ''dewatering'' and drying circuits can be avoided; and (4) the overall cost of cleaning may be reduced once the technologies are applied in full-scale commercial production.

However, several disadvantages may limit the usefulness of these techniques: (1) they cannot remove the organic sulfur from raw coal, as chemical coal cleaning attempts to do (see the chapter Chemical Processes); (2) the initial capital and operating and maintenance cost projections may not indicate the final cost of cleaning coal in full-scale operations; and (3) some processes may present problems of waste disposal and polluting emissions.

Some of the advanced physical coal cleaning technologies appear to be suited to commercial use. They are either in the experimental (laboratory) or pilot/demonstration-plant phase of development. Most

ADVANCED PHYSICAL COAL CLEANING TECHNOLOGIES

Developer Process	Year Work Started	Sulfur Reduction Achieved* (Best test case)
Advanced Energy Dynamics Electrostatic Separation	1979	37 to 68% total sulfur removal
Helix Corporation Clean Pellet Fuel	1975	2.12% sulfur raw coal; clean pellet emits 0.34 lbs sulfur per million Btu
Iowa State University Oil Agglomeration	1974	88% pyritic sulfur removal

Oak Ridge National
Laboratory
  High Gradient Magnetic
Separation
  Open Gradient Magnetic
Separation
	1976	60% pyritic sulfur remov-
		al-High Gradient Magnet-
		ic Separation (HGMS)

Otisca Industries, Ltd. Otisca Process	1972	70% pyritic sulfur remov- al (63% of the total sul- fur removed)

Ash Reduction Achieved	Technical Approach	Markets	Costs
51 to 59% ash removal	Finely pulverized coal fed into modern iron ore cleaning modules that separate coal from wastes because of their differences in electrical conductivity	Utilities, industrial coal users	Company projects that electrostatic cleaning is 32 to 77% cheaper than "conventional" coal cleaning technologies
Not available	Raw coal combined with limestone to form burnable pellet. Pellet burned directly or gasified	Industrial boilers, furnaces in hospitals and prisons	Raw Coal: $1.20/million Btu; Clean Pellet Fuel: $2.00/million Btu
63% ash removal	Fine coal/water slurry combined with fuel oil to encourage separation of coal and wastes	Utilities, industrial coal users (that can burn a coal/oil mixture)	For theoretical 200 ton per hour plant, capital and fixed costs estimated at $1.50 per ton of raw coal processed (1974 dollars)
50-60% ash removal (HGMS)	Fine coal cleaned with dry, magnetic equipment that removes impurities from coal	Utilities, industrial coal users	For a 200 ton per hour HGMS plant: operating costs of $3.78 per ton of clean coal, with a capital investment of $23 million; comparable "conventional" plant would have operating costs of $4.97 per ton of cleaned coal and a capital investment of $16 million
87% ash removal	Heavy organic liquid (CCl_3F) used as cleaning bath instead of magnetite/water mixture to clean coal by gravity separation	Utilities, industrial coal users, metallurgical coal users, synfuel producers	For a projected 400 ton per hour plant: Capital costs of $6 million; Direct Operating Costs of $1.45 per ton of raw coal; Total Costs of $3.65 per ton of raw coal (1979 dollars)

of their developers were reluctant to predict exactly when they would be applied commercially. However, estimates of the period of time needed for commercialization range from one year or less to over five years. Of the technologies profiled, the Otisca Process is the closest to being commercialized—a demonstration plant is operating at the site of a utility-owned mine in Ohio. By contrast, the "oil agglomeration" process has so far only been tested under laboratory conditions at Iowa State University.

Five advanced physical coal cleaning technologies are profiled in this report: Clean Pellet Fuel, oil agglomeration, magnetic separation, the Otisca Process, and two miscellaneous techniques (electrostatic separation and the Kintyre Process). These technologies have been developed in several settings, including corporate, academic and government research laboratories. The sample represents some of the most sophisticated research in physical coal cleaning, and covers most of the novel technical approaches to cleaning coal by physical means. In three cases, other U.S. research groups have been investigating similar techniques. The groups chosen for this report either are further along than other researchers in developing the processes or offer greater promise in their technical achievements.

The preceding table summarizes the advanced physical technologies profiled.

The advanced physical coal cleaning technologies were developed during the early and middle 1970s. With the exception of Clean Pellet Fuel, these processes remove over half the ash and inorganic sulfur from the raw coal. Clean Pellet Fuel emits very little sulfur dioxide when burned, but leaves substantial quantities of "bottom ash" in the boilers. This fuel differs from the others (see Helix Technology Corporation profile). With the exception of the Kintyre Process, the remaining fuels are based on concepts that are closer to the conventional approaches to coal cleaning (see profiles of Other Technologies).

According to the organizations profiled, the markets for the advanced physical processes range from large utility boilers to small units located in hospitals, schools and prisons. Several technologies can provide fuel for users with special needs, such as the synfuels industry and utility or industrial boilers that burn a mixture of coal and oil.

Capital investment and operating cost estimates for the processes were provided by their developers. Although comparisons are difficult to make, the developers claim that their technologies can save more money than most conventional coal cleaning techniques. It is important to note that the advanced physical coal cleaning methods, like any new technology, are still untested in day-to-day operations in commercial settings, where unexpected difficulties may increase costs and decrease efficiency.

Chapter 6

Introduction to Chemical Coal Cleaning

On the forefront of the physical coal cleaning technologies today are innovations such as froth flotation units, multiple screening devices and magnetic separators; all of which use the difference between the physical properties of coal and its impurities to separate the two. Chemical coal cleaning (CCC) uses more sophisticated chemical processes to remove sulfur from coal, including the tiniest particles of pyritic sulfur and some of the organic sulfur which are beyond the removal ability of any physical coal cleaning process.

Present CCC technology can remove almost 100 percent of the pyritic sulfur from most coals. Total sulfur removal by various methods of CCC, including the removal of organic sulfur, now ranges from 50 to nearly 90 percent. Presently, CCC research and development is aimed at attaining the cost-effective removal of at least 90 percent of the sulfur in coal, thereby allowing new power plants to burn coal without scrubbers.[1]

According to organizations developing these processes, CCC is six to ten years away from commercial application.

In 1980, INFORM selected nine government and corporate organizations that were developing the most advanced chemical coal cleaning processes to remove pyritic and organic sulfur, and whose programs were expected to continue. Three of the programs studied, the Ames Laboratory, the Jet Propulsion Laboratory and the Pittsburgh Mining Technology Center, were government projects, and the rest were corporate. All of these programs, except ARCO, depend now, or did depend on government research funds from the Department of Energy (DOE).

Present Status of Research and Development

Most of the research into CCC began before 1977 when emission standards for utility boilers were less stringent than they are now. While government and corporate organizations have developed processes, designed plants and estimated their potential problems, benefits and costs, not a single CCC process has been developed to the full pilot plant stage; although eight of the developers claim to be ready for pilot plant programs which would cost from $5 to $10 million.

As a result of the 1977 amendments to the Clean Air Act and the new Administration's emphasis on private research and development, only two chemical coal cleaning methods, GE and TRW's gravimelt— the only processes that now promise to desulfurize coals to a degree which entirely eliminates the need for stack gas scrubbing at new utility plants—are now being funded by the DOE. Two other CCC projects studied, ARCO's and Kennecott's, are continuing with their own funds.[2]

The DOE budget for research into CCC decreased from $4.4 million in 1980 to $3.0 million in 1981, and further decreases are expected in 1982. The 1982 $1 million OMB appropriation for CCC has not yet been approved by both houses in Congress.[3]

Process Description

For the purposes of this discussion, chemical coal cleaning will include those methods of chemical coal cleaning that use regenerable industrial chemicals, in a procedure in which coal remains in the solid form throughout processing, and in which solid desulfurized coal is the only fuel product. Not included in this discussion are methods of coal treatment that partially or completely convert solid coal into another form. All of the methods of chemical coal cleaning described below have the potential to extract organic as well as pyritic sulfur.

In most chemical coal cleaning processes, coal is usually crushed to

-14 mesh.[*][4] Chemical desulfurization techniques work best on coal which has also been physically cleaned (and crushed to fine or middling size) as large amounts of coarse pyrite actually inhibit the chemical reactions which remove organic sulfur.[5] Coal with pyrite and ash forming minerals also tend to use up more reagent in CCC than does cleaned coal, adding an unnecessary cost.[6] For these reasons, economic estimates of CCC assume that the coal feed is physically washed. This coal is mixed with water or a water and chemical reagent solution to form a slurry. The slurry and any additional chemical reagents are fed into a reactor vessel, which is sealed, heated and/or pressurized.[†] Under these conditions, much of the sulfur in the coal goes into solution with the liquid reagents.

All nine of the major chemical coal cleaning technologies presented in this report utilize either "oxidative" or "caustic" leaching to remove sulfur. In "oxidative" leaching, individual organic sulfur atoms are oxidized to form sulfur oxides (SO_x) or sulfates (SO_3), chemical groups that are easier to "cleave" or break from the rest of the coal molecular structure than unoxidized sulfur. "Caustic" leaching utilizes strongly basic substances such as lye, that actually displace organic sulfur atoms from the coal molecular matrix.

Dr. Robert Meyers of TRW has noted that a CCC reagent must have four specific attributes: 1) it must react very selectively with a coal's sulfur and not with the coal itself; 2) it must be highly recoverable from the coal after the desulfurization process; 3) it must be regenerable, so that most of it can be chemically processed back into its original form; and 4) it must be inexpensive, as a small amount will always be left on the coal and have to be replaced.[7] Satisfactory reagents are in use that remove pyritic sulfur and some organic sulfur.

Coal emerges from the reactor vessel impregnated with the sulfur-containing reagent fluid. It is washed with a water solution to remove the reagent fluid and is pressed into a "cake" and dried. The used slurry fluid is then passed over lime, which separates the water from the used reagent. This water is recycled back into the process and the used reagent is processed further to separate a sulfur containing waste from

---

*This is much smaller than the size of particles treated in a conventional coal cleaning plant, where the smallest particles are generally 28 mesh.

†The GE process is an exception to this rule as it works on a dry coal, not on a slurry, and operates at ambient temperatures.

Company and Primary Business	Name of Process	Research Began
Ames Laboratory Government-funded research laboratory	Ames Oxydesulfurization (alkaline oxydesulfuriza-tion)	1974
Atlantic Richfield Company Diversified natural resource company	ARCO desulfurization	1976
Battelle Memorial Institute Diversified natural resource company	Battelle hydrothermal process	late 1960's
General Electric Electrical equipment, mater-ials, natural resources, ser-vices and transportation equipment	GE microwave process	1975
Hydrocarbon Research Institute Research and engineering firm	Kennecott oxydesul-furization process	1979
Jet Propulsion Laboratory Government-sponsored space and energy lab-oratory	Chlorinolysis process JPL I uses chlorine gas JPL II uses water instead of chlorine	1975
Pittsburgh Energy Tech-nology Center National energy research center	PETC oxydesulfurization process	1970
Research-Cottrell Energy and pollution control equipment	KVB - Guth process	1970
TRW, Inc. Research, Engineering and Manufacturing Firm	Gravimelt	1979

*Percentages are approximate and may vary within a few percentage points

Funds Available for Research	Sulfur Removed(%)			Ash Removed(%)	Btu Loss(%)
	Pyritic	Organic	Total		
$500,000 as of June 1980. Half from DOE and half from State of Iowa	95	30	70*	20	5
Estimated total exceeds $7 million. $1 million from EPRI, $6 million from ARCO	98*	35	70-80	NA	10
Several million dollars. Approximately 90% from Battelle and 10% from EPA	95	25-40	NA	NA	NA
Over $1 million. 80% by DOE, EPA, and NSF, and 20% by GE	Method 3 95*	60*	80*	NA	NA
Several million dollars invested by Kennecott. HRI also spent several man years in process engineering work. Recent government investment for testing by TRW (DOE)	Oxygen No. I 90 / No.II 90 Ammonia	15-25 / 30-40	70 / -	none / -	10 / -
Approximately $1.4 million, 85% from DOE and remainder from Bureau of Mines and JPL			65* 70*	4	10
$1,285,000-- $400,000 in 1979 and $885,000 in 1980 from DOE	90*	40*	70	10	10
$665,000 - KVB provided $100,000 and Research Cottrell the remainder. May 1980, DOE grant of $415,000	95*	40*	50-70	NA	NA
$450,000--$307,000 from DOE contract awarded in May 1980	-	-	80*	50-95	5

as much of the reagent as possible. The purified reagent is then used again.

The reagent regeneration step produces chemical coal cleaning's principle by-product, gypsum. The wash water from all the processes profiled must be cleaned and reused to ensure economic viability, so the by-product gypsum is left relatively dry. Because the desulfurization step also removes pyrite (FeS_2), iron oxides are formed and end up in the gypsum as rust, making it brown in color. This gypsum also contains significant quantities of possibly toxic trace metals removed from the coal during desulfurization.

At present, it appears that techniques of caustic leaching have the potential to remove a higher percentage of sulfur than methods of oxidative leaching. Higher percentages of sulfur removal are also realized with both techniques when they are applied to coals with a lower organic and higher pyritic sulfur content.

This general approach to chemical desulfurization is the same for all of the processes profiled. Of course, variations on each step of the process outlined above occur within the individual techniques. For example, the GE process works on dry coal, not on a slurry; and the ARCO process uses a non-regenerable reagent.

Present Chemical Coal Cleaning Capabilities and Applications

As noted earlier, chemical coal cleaning can remove both the pyritic and organic sulfur compounds in coal, whereas physical coal cleaning can extract most of the pyritic sulfur, but none of the organic sulfur. The organic sulfur content of U.S. coals, however, does account for a substantial portion of the sulfur in coal:

Average Sulfur Content of 455 U.S. Coal Beds[8]
Representing Principal Coal Producing Regions

	Sulfur Content By Weight (%)
Total Sulfur:	3.02
Pyritic Sulfur (mineral):	1.91
Organic Sulfur (By Difference):	1.11

According to INFORM's sample, the most advanced physical coal cleaning techniques remove at most 75 percent of this pyritic sulfur (or 47 percent of the total sulfur*). Thus, an "average" U.S. coal, physically cleaned by the best available practice, would still contain 1.4 percent total sulfur. (Some Appalachian and midwestern coals would nearly double this "average" figure.)

All of the CCC processes examined in this report have been proven to be highly effective in removing almost 100 percent of the finely dispersed pyrite left in coal after physical cleaning. This pyrite is not chemically bonded to coal molecules, but is present in such microscopically sized pores, that even in very finely crushed coal, water cannot wash it away. Chemical coal cleaning reagents can react with this pyrite, making it water soluble so that it can be easily removed. Chemical coal cleaning tests have shown removal capabilities from 50 to 88 percent of the total sulfur, including organic sulfur, in high sulfur coals. Three of the processes profiled, ARCO, GE and TRW's Gravimelt, have repeatedly achieved 80 percent total sulfur reductions in recent tests.[9] While these tests on high-sulfur eastern and midwestern coals are limited to bench scale experiments, they indicate the possibility of significantly desulfurizing U.S. coals.

Methods of CCC presently under development can also remove significant quantities of trace elements and some nitrogen in addition to the sulfur. This reduces the amount of possibly dangerous trace elements and nitrogen oxides emissions released during combustion. However, some chemical coal cleaning processes do not remove ash as effectively as does physical coal cleaning, and therefore CCC works best on coal that has been partially cleaned, with the coarse pyrite crushed and the coal reduced to fine and middling size.

Chemical coal cleaning plants could be built using existing flow sheets and almost entirely with proven industrial parts. The only new component that would have to be designed would be the desulfurization vessels. All of the other components, slurry pumps, coal pulverizers, driers, thickeners and filters, are proven technologies.

The environmental and economic importance of "clean" coal is a result of EPA's Revised New Sources Performance Standards (RNSPS) for utility boilers, which requires the removal of at least 90 percent of the sulfur present in a coal except where emissions are held to 0.6

---

*The Homer City preparation plant began achieving this sulfur reduction in March of 1981.

pounds of sulfur dioxide or less; and then, 70 percent of the sulfur originally present in coal must be removed.

Data on two representative high sulfur coals provided by DOE illustrates the possible significance of CCC in meeting this EPA regulation:

COAL TYPE[10,11]

	Kentucky No. 11		Illinois No. 6	
	Total Sulfur (%)	$SO_2/10^6$ Btu(lbs)	Total Sulfur (%)	$SO_2/10^6$ Btu(lbs)
Raw Coal (typical) (1):	4.40	6.9	3.00	4.8
Physical Cleaning Only (2) (typical):	3.00	4.5	2.54	3.8
Sulfur Reduction(%):	31	–	15	–
Combined Physical and Chemical Cleaning (3) (experimental):	.53	0.80	.36	.56
Sulfur Reduction(%):	87	–	88	–

In terms of potential sulfur dioxide emissions and stack gas scrubbing necessary to comply with the New Source Performance Standards, the Illinois No. 6 coal conditions compare as follows:

Coal Condition	$SO_2/10^6$ Btu Which Must Be Scrubbed	FGD Sludge Produced In[12] Scrubbing(Tons/Year)
Raw:	3.38	140,000
Physically Cleaned:	3.24	132,000
Combined Physically-Chemically Cleaned:	None	None

These figures apply to a new 500 Mw plant, with a thermal efficiency of 33 percent and an 85 percent capacity factor.

In addition to providing a cleaned fuel for utility boilers, the chemical coal cleaning process developers also see small coal-fired boilers, generally termed industrial boilers, as a potential market. For such boilers, which need not meet the RNSPS, chemical coal cleaning could be a more economical method than scrubbers of meeting emission requirements. Industrial users would thereby avoid the necessity of finding the space required by scrubbers, and could avoid a scrubber's waste disposal problems. Furthermore, industrial boilers, unlike most utility boilers, are often turned off and on intermittently, which can cause problems in the use of an FGD system (see INFORM's *The Scrubber Strategy*), but would have no effect on the use of chemically cleaned coal.

In 1977, industrial boilers used approximately ten percent of all U.S. coal. A DOE spokesman estimates that growth in the industrial marketplace and conversions of industrial boilers to coal in the next decade could make the market for CCC as much as 200 million tons per year.

Problems of Chemical Coal Cleaning Technology

Chemical coal cleaning does create environmental hazards whose impacts must be minimized. Coal storage and handling presents a dust and particulate problem similar to that of physical cleaning but one that is aggravated by the high volume of very fine sizes handled in chemical coal cleaning plants.[13] Though sulfur is usually carried away from CCC reactors in liquid solution, some research has suggested that sulfur-containing gas is also produced by the desulfurization reactions.[14] This gas, probably mostly sulfur dioxide, although other gases are also present, would escape into the atmosphere when the sealed reactor is vented. In any commercial CCC plant, such gases might have to be scrubbed to meet EPA regulations—an ironic dilemma for a technology which was created as an alternative to FGD. Some developers claim desulfurization conditions can be adjusted to minimize these "off-gases."[15,16,17] However, more research is needed to fully assess the potential environmental impact.

Eight of the nine developers profiled anticipate marketing their "abatement" gypsum (ARCO is the exception).[18] However, the market possibilities for this product are unclear. The wallboard industry utilizes 73 percent of all U.S. gypsum, but the brown CCC product would not be an acceptable substitute for white and inexpensive natural gypsum. The cement industry also utilizes gypsum, but it is uncertain whether that industry could absorb the quantities of gypsum produced in large volume coal processing.

The alternative to marketing by-product gypsum is to dispose of it in a landfill where the leaching of trace metals would have to be controlled. However, the circumstances under which CCC waste leaches have yet to be established, and dangers of disposal will only become clear after complete pH and compositional analyses have been done on large quantities of actual refuse, i.e., after scaled-up research.

Cost Estimates

The primary consideration affecting chemical coal cleaning development planning is the cost effectiveness of the technology. Chemical coal cleaning will cost more than physical cleaning. Basic equipment costs will be higher and in addition, chemical reagents, recycling, and post-treatment requirements can be expected to add further expense.

To be cost effective, therefore, chemical cleaning must provide considerably greater benefit, in terms of the ultimate cost of energy, than physical cleaning. In other words, the ''cost'' of chemical cleaning cannot exceed its ''worth'' to the user.

Preliminary estimates of the costs of the nine methods studied range from $0.61/10^6 Btu for scrubbers, and $3.00 to $4.00/10^6 Btu for liquefaction and gasificaction.[19,20,21,22] Chemical coal cleaning is 10 to 30 percent more expensive than scrubbers and at least three to four times less expensive than coal liquefaction and gasification.

However, this economic comparison does not take into account the benefits that CCC could provide in addition to eliminating the need for flue gas scrubbing at a power plant. An unpublished DOE report calculated the potential worth of CCC as follows:

1. The elimination of flue gas scrubbing in a modern 500 Mw power plant could be worth as much as $30 per ton of coal;

2. Further benefits realized in transportation and boiler operations could be worth another $20.00 per ton;*

3. On this basis, the potential value of a chemically cleaned "supercoal," would be around $80 per ton (assuming input coal at $30 per ton). †[23]

According to these DOE figures, to be commercially viable, CCC must remove enough sulfur to eliminate scrubbers, at a cost of less than $50 per ton. Preliminary cost estimate by Bechtel, Versar and others vary widely, from as low as $15 to $20 per ton to as much as $50 to $60 per ton.[24] Thus, according to the DOE, these cost estimates "do not appear to present intrinsic processing cost barriers which would preclude commercialization."[25]

However, as the DOE report mentions, these cost estimates have been derived from extrapolations of laboratory-scale experiments and should not, therefore, be viewed as anything other than rough order-of-magnitude calculations. Dependable estimates will require experience with small-scale prototypic hardware and could be lower or higher than these present estimates.

The costs of CCC vary because the specific process steps which are responsible for most of the cost of CCC differ from process to process. However, some general observations can be made by dividing the costs of chemical cleaning into electricity, reagents, and lime.

The largest cost for most CCC processes is the electricity used to heat and/or pressurize the desulfurization reactors, to operate the slurry pumps (which move coal/water/reagent mixtures through the CCC system), and to dry the coal.[26,27] These electricity costs equal approximately 40 percent of the annual operating budget of every chemical coal cleaning process profiled.[28,29]

However, techniques to decrease electricity usage are being designed

---

*The report notes, however, that accurate assessments of these benefits (as with physical coal cleaning) require detailed knowledge of specific site conditions, i.e., size, age, design of boilers, how often the plant is used, location of plant, and other local considerations.

†In 1981 dollars, a chemically cleaned "supercoal" would be worth about $3.10/10^6 Btu, on this basis which is equivalent to fuel oil at $17.86 a barrel—about half today's oil cost (roughly $6.00/10^6 Btu).

Estimates For A Theoretical System
Costs (1980)*

(8,000 ton per day plant)

Company	Total Capital	Annual Operating	Size of Latest Test
Ames Laboratory	$277,000,000	$21,240,000 ($1.48/MMBtu)	1 liter autoclave on 10-20 coals
Atlantic Richfield Company	$191,000,000+	$58,400,000+ ($0.92/MMBtu)	200 lb/hr pilot plant
Battelle Memorial	$168,100,000**	$74,800,000** ($1.31/MMBtu)	1/3 ton per day mini-plant
General Electric	(Method 3) $102,000,000**	$35,900,000** (+$0.61/MMBtu)	Batch tests of 100-200 grams/minute
Hydrocarbon Research Institute	No. 1 (oxygen) $255,000,000	$18,480,000 ($1.41/MMBtu)	Batch tests (no size given)
Jet Propulsion Laboratory	JPL I: $269,000,000 (chlorinolysis process) (uses water solvent)	$29,770,000 ($1.76/MMBtu)	2 kilograms/hour continuous mini pilot plant
Pittsburgh Energy Technology	$301,000,000	$18,840,000 (+$1.68/MMBtu)	Stirred tank reactor (no size given)
Research-Cottrell	$208,000,000	$32,110,000 ($1.22/MMBtu)	100 gram, 10 mesh coal samples in bench and continuous batch reactors
TRW, Inc.	$73,460,000§	$52,300,000§ ($/MMBtu unavailable)	1 to 2-gallon batch reactors

*All estimates given, unless starred, are by Bechtel National, Inc. Bechtel's 1980 estimates include costs for physical coal cleaning prior to treatment
+Estimates by ARCO and EPRI. Costs exclude physical coal cleaning
**Estimates by Versar, Inc. Costs include physical coal cleaning
§Estimates by TRW. Costs exclude physical coal cleaning

into process flow sheets. For example, a possible alternative to electrically heated reactors is to burn some of the cleaned coal. However, the cost of using as much as 7 or 8 percent of the product tonnage may not be less than the cost of the electricity. Centrifuges to partially dewater coal would use less electricity than thermal dryers, but centrifuges cannot always bring washed coal to a moisture level acceptable for firing in a pulverized-fuel boiler.

The most significant cost of CCC after electricity is the replacement cost of used reagents. This cost is specific to each process and to the kind of reagent used, and in some cases, it can exceed the cost of electricity. While recovery and regeneration levels may be improved, 100 percent reagent recovery will never be obtained in the processes now under study.[30] Reagent replacement cost is expected to be a major consideration in the JPL 1, Guth, Battelle, ammoniacal Kennecott, and Gravimelt processes.[31,32,33]

Included in this reagent category are process gases which are used as reagents. Oxidative processes which occur in oxygen are sometimes called oxydesulfurization techniques. Oxydesulfurization processes will require large oxygen manufacturing plants adjacent to CCC facilities.[34] Since this oxygen will be manufactured on site, the cost is not determined by commercial markets as are those of regularly purchased chemicals. Still, oxygen is a major reagent cost in the Kennecott, Ames, and ARCO processes.[35,36] In some cases, the presence of large quantites of oxygen presents a danger of combustion when used with a highly combustible fuel such as coal. However, oxygen plants have been operated safely at the site of other industrial facilities.

The cost of lime accounts for an average of 20 percent of the annual operating budget of each process.[37] This cost is expected to remain relatively stable.

The Future of Chemical Coal Cleaning Technology

Although chemical coal cleaning will benefit from further research in several areas, improvement in its ability to remove organic sulfur to consistently achieve a.90 percent sulfur reduction is essential if CCC is ever to become a commercial, viable technology.

Lack of sufficient theoretical knowledge about the molecular structure of coal is a key factor which now limits the capacity of CCC to remove organic sulfur. Organic sulfur appears in many different forms within coal's complex molecular structure, not all of which are known. The problem encountered in CCC is that the different forms of organic sulfur cannot be separated from coal under the same reaction conditions.[38]

Organic sulfur is present in coal in two principle forms, chain and ring sulfur. Some researchers suggest that it is almost impossible to break down ring sulfur, and that whatever sulfur is being removed by CCC is of the chain type.[39] There is growing evidence that roughly half of the organic sulfur in coal is of the ring type.[40] For these reasons it appears that there may be an upper limit to organic sulfur removal by both caustic and oxidative leaching. Coals with a high percentage of organic sulfur may not even be able to reach as much as 50 percent desulfurization.[41,42,43,44]

A related problem which troubled many CCC developers profiled in this report is the inadequacy of the standardized American Standard Testing Association (ASTM) procedure for measuring organic sulfur. The ASTM test is not a direct determination of organic sulfur content but a determination by difference (total sulfur minus pyritic sulfur) and it does not provide researchers with sufficiently accurate data. Consequently, several programs are now under way to develop a direct method for determining the amount of organic sulfur in coal.[45]

Problems also surround the drying and transportation of the CCC product, which is usually P.C. grind* or finer. Drying this product presents many of the same problems as drying fines from conventional preparation plants, and some research is now focused on performing CCC on a coarser grind of coal.[46] Although chemically cleaned coal would probably be pressed into cakes or pelletized, the difficulties of shipping large quantities of such coal are still unknown.

During fiscal year 1982 (which began October 1, 1981), DOE will be analyzing the sulfur removal capabilities and cost projections for the two CCC processes it is now sponsoring (GE and TRW's gravimelt) to determine which, if any, should be sponsored further. Bench scale tests are being continued to determine the relative capabilities of these processes. According to DOE, at the present time TRW's fused salt process appears to provide the most complete desulfurization. DOE anticipates that one or both of the candidate processes will be taken into the next phase, the objective of which will be to resolve critical process

---

*P.C. grind or "pulverized coal" is defined to be 70 to 80 percent 200 mesh.

design uncertainties. Process steps which present the highest cost risks (e.g., filtering, reaction kinetics, material recovery, etc.) will be studied with respect to plant design, product quality, and material requirements.[47]

When this work is completed, DOE states that the design of a prototype, small scale (5 tons per day) "process feasibility unit" (PFU) will be undertaken. The goal of this project, according to Agency officials, is to demonstrate the commercial viability of CCC in order to eliminate the need for flue gas scrubbing of selected high-sulfur U.S. coal in their respective utility markets.[48]

Part IV

Corporate Profiles

Architect/Engineering/
Construction Firms

ALLEN & GARCIA COMPANY
332 South Michigan Avenue
Chicago, IL 60604

Allen & Garcia Company of Chicago, Illinois, was founded in 1911 as a private business. On July 1, 1979, the company was acquired by Simon Engineering Corporation of Stockport, Great Britain, a worldwide engineering and construction firm. Allen & Garcia designs and builds a full range of material-handling and coal-preparation systems, varying in size from small feasibility studies and minor plant modifications to major building projects costing up to $50 million. Other engineering and construction projects have included raw-coal storage and blending, unit-train loading systems, mining shafts, railroads, dams, roads, and auxiliary mine facilities.

Allen & Garcia does not manufacture and sell its own coal-preparation equipment. Since the company does not manufacture preparation plant equipment, it claims to take an objective approach to construction projects, without bias toward any type of component or coal cleaning technology. Owen Brumbaugh, Allen & Garcia's Vice President of Engineering, sees the company as being open to whatever technology is most appropriate for a given application, and capable of designing and building any type of coal cleaning system.

No financial information was given by Allen & Garcia conerning its revenues or profits.

Coal Cleaning Plants

Total number of plants constructed
or retrofitted (through 1980): 226

> To prepare steam-grade coal: 2/3 of total
> To prepare metallurgical-grade coal: 1/3 of total

Number of contracts:

> 1970 to 1975: NA
> 1976 to 1980: NA

Plant profiled:

> Western Slope Carbon Coal Company Preparation Plant,
> Northwest Coal Corporation

Manufactures preparation plant equipment

DRAVO CORPORATION
One Oliver Plaza
Pittsburgh, PA 15222

Dravo Corporation is descended from a business started in Pittsburgh, Pennsylvania, by F. R. and R. M. Dravo in 1891. For the past ten years, Dravo has been engaged in the design, engineering and construction of coal preparation plants. Dravo is a highly diversified concern, providing goods and services worldwide. Its customers include the chemical, petrochemical, metals and minerals, and utilities industries. Dravo sells chemical and mineral raw materials, transportation equipment and services, plant equipment and process-plant engineering, and construction project management and technical services.

Coal preparation plants are designed and constructed by Dravo Engineers and Constructors, a subsidiary of the parent corporation that accounts for 50 to 60 percent of its revenues. This subsidiary can also design and build any of the other systems used by the coal industry, from the coal mine to the power plant.

Since Dravo Engineers and Constructors is a large, diversified enterprise, it can, along with the parent corporate structure, provide considerable engineering expertise and capital resources for a coal operation's preparation plant activities. Dravo's engineering organization has a "matrix-oriented" structure, where people from different divisions within the corporation can lend their particular expertise to projects in other divisions. According to Dravo, its areas of expertise in coal preparation plant design and construction are high-percentage Btu-recovery circuits, fine-coal cleaning processes and dewatering, and the dewatering of fine wastes.

In 1980 Dravo Corporation had revenues of $1.113 billion and a pretax income of $35.9 million. The company would not say what percentage of these revenues or of this net income was generated by work on coal preparation plants.

Coal Cleaning Plants

Total number of plants constructed
or retrofitted (through 1980): 10

> To prepare steam-grade coal: 5
> To prepare metallurgical-grade coal: 5

Number of contracts:

> 1970 to 1980: 10

Plants profiled:

> Warwick Preparation Plant,
> Duquesne Light Company

> Marrowbone Mine Preparation Plant,
> A.T. Massey Coal Company

> Helper Preparation Plant,
> American Electric Power Company

Does not manufacture preparation plant equipment

HEYL & PATTERSON, INCORPORATED
250 Park West Drive
P.O. Box 36
Pittsburgh, PA 15230

Heyl & Patterson, Inc. (H&P), of Pittsburgh, Pennsylvania, is a private corporation established in 1887. The firm, which has been designing, engineering and constructing coal-preparation equipment since its founding, sees itself as a "pioneer" in the field and a "training ground" for others in the coal-preparation industry. H&P recently became involved in several related fields, including potash and other mineral beneficiation, and the manufacture of material-handling equipment and thermal-dryer systems.

Heyl & Patterson builds coal preparation plants and manufactures the equipment used in these plants. As a preparation plant equipment manufacturer, H&P can sell equipment to its competitors even if it is not awarded a contract to build a given plant.

H&P notes that it was one of the first firms to develop some of the equipment used in coal preparation, including classifying cyclones, heavy-media cyclones, and fluid-bed dryers. The company also took part in the development of methods to clean fine coal particles.

Heyl & Patterson would not provide figures on its 1980 revenue or net income. As a privately held corporation, its financial statistics are not available to the public.

Coal Cleaning Plants

Total number of plants constructed
or retrofitted (through 1980): 50(+)

     To prepare steam-grade coal: 50%
     To prepare metallurgical-grade coal: 50%

Number of contracts:

     1970 to 1975: 20
     1976 to 1980: 10

Plant profiled:

     Homer City Preparation Plant,
     Pennsylvania Electric Company/
     New York State Electric & Gas Corporation

Manufactures preparation plant equipment

KAISER ENGINEERS, INCORPORATED
300 Lakeside Drive
P.O. Box 23210
Oakland, CA 94623

Kaiser Engineers of Oakland, California, is one of the largest companies in the engineering and construction industry. It was acquired by Raymond International, Inc., in May 1977. A highly diversified company, Kaiser is also active in the fields of hydroelectric power, rapid transit, cement production, and metals processing.

The company has not constructed a coal preparation plant in the United States to date, although it has built two plants and retrofitted seven others in Canada since the early 1970s. However, Kaiser has done a design and feasibility study for one preparation plant a month from 1974 through 1981.* Kaiser states that its lack of construction business can be traced to two factors: (1) utilities and coal companies are uncertain about the direction of government environmental regulations, and are therefore reluctant to commit themselves to plant construction; and (2) the government has not yet made a definite commitment to increasing coal use, which has also made potential customers less willing to build plants at this time.

Kaiser Engineers has an Environmental Control Department with both air and water quality divisions. Unlike most preparation plant A/E/C firms, Kaiser will complete the permit applications needed before the construction of a plant if a client so desires. If the client has filled out its environmental permit requests, Kaiser will check the requests and approach the appropriate federal agency with the client for permit approval.

---

*Of these 70 to 75 studies, three quarters were designs for plants that would produce steam coal, and one quarter of the designs were for plants that would prepare a metallurgical-grade coal.

Kaiser Engineers also has a unique "Financing Group" that can help a customer put together a financing package for a preparation plant, although Kaiser will not actually finance the plant for the customer. Further, Kaiser offers legal counsel to its clients and a technical support staff with experience in land acquisition.

In addition to designing and building preparation plants, Kaiser Engineers manages coal-research programs developed and funded by other organizations. Kaiser managed over 15 preparation plant research projects for the Electric Power Research Institute (EPRI). At present, Kaiser has three coal cleaning assignments from EPRI: to study heavy-media cleaning, to study the dewatering and drying of fine coal particles, and to oversee the engineering work at the EPRI Homer City Preparation Plant Test Facility.

Kaiser Engineers will also develop an organization's coal cleaning research and development program. In 1977 the company worked with EPRI to develop its five-year R&D program. One of Kaiser's suggestions was that EPRI build a preparation plant test facility. Such a plant, the EPRI Homer City Preparation Plant Test Facility mentioned above, is now in the start-up phase (see also EPRI profile).

The following tables provide financial data on Kaiser Engineers and its parent company for the fiscal years 1979 and 1980.

Coal Cleaning Plants

Total number of plants constructed
or retrofitted (through 1980): 10

>   To prepare steam-grade coal: 1
>   To prepare metallurgical-grade coal: 2

Number of contracts:

>   1970 to 1980: 10

Plant profiled:

>   Homer City Preparation Plant,
>   Pennsylvania Electric Company/
>   New York State Electric & Gas Corporation*

Does not manufacture preparation plant equipment

---

*General Public Utilities Corporation chose Kaiser Engineers to help it solve the operating problems at Pennsylvania Electric Company's Homer City Preparation Plant. Kaiser came up with design solutions to two of the problems at this plant: the cleaning of fine coal particles in heavy-media equipment and the recovery of magnetite from these vessels. These adjustments will now be made by Kaiser. (For additional information on the Homer City plant, see Pennsylvania Electric Company profile.)

MCNALLY PITTSBURG MANUFACTURING COMPANY
101 East Fourth Street
P.O. Box 651
Pittsburg, KS 66762

McNally Pittsburg Manufacturing Corporation of Pittsburg, Kansas, was launched in 1889 as a boiler works that repaired locomotive engines. The company became involved in coal preparation in the 1920s when coal mining was still a thriving business in the area. Since its beginnings, McNally Pittsburg Manufacturing Corporation has grown into the six-member McNally Group.* The McNally Group is one of the largest worldwide conglomerates offering services to the mining industry. The Group has design, engineering and construction capabilities in the hard-rock-minerals and coal fields, and it offers calcining kilns, crushing equipment, conveying systems, large steel-fabrication equipment, systems for high-density slurry recovery, and coal preparation plants and coal-handling systems.

McNally Pittsburg's Kansas and Ohio divisions have specialized in the design and construction of coal preparation plants and handling systems. The Kennedy Van Saun Corporation in Danville, Pennsylvania, and McNally Utah in Salt Lake City, Utah, also provide these services. Each of the four members of the McNally Group having preparation plant experience generally does work in its own territory. The Kansas Division was interviewed for this study.

Thomas McNally is said to have imported the first coal-washing technology from Europe. McNally designs and manufactures a great deal of preparation plant equipment such as Baum-type jigs, dense-media vessels, heavy-media cyclones, water-cleaning cyclones, classifying cyclones, centrifugal dryers and fluid-bed dryers. "We are not married to our equipment but a customer may benefit from using it," asserts Ernest Draeger, Chief Process Engineer with McNally's Kansas Division. He points out that "our customers can go to one company for guarantees on preparation plant operation as well as for the smooth running of components."

---

*The members of the McNally Group are the Kansas and Ohio divisions of McNally Pittsburg Manufacturing Corporation, the Kennedy Van Saun Corporation, McNally Utah, the McNally Mountain States Steel Company, and Marconaflo.

McNally Pittsburg, a privately held company, declined to provide figures on its 1980 revenues or net income.

Coal Cleaning Plants

Total number of plants constructed
or retrofitted (through 1980): about 500

    To prepare steam-grade coal: "majority"
    To prepare metallurgical-grade coal: NA

Number of contracts:

    1970 to 1975: 53 (plus 15 handling systems)
    1976 to 1980: NA

Plants profiled:

    Rum Creek Preparation Plant,
    Pittston Company

    No. 2 Preparation Plant,
    Monterey Coal Company

Manufactures preparation plant equipment

ROBERTS & SCHAEFER COMPANY
120 South Riverside Plaza
Chicago, IL 60606

Roberts & Schaefer Company (R&S) of Chicago, Illinois, has been engaged in the design, engineering and construction of coal preparation and mineral beneficiation plants, and related material-handling facilities, since 1903. In 1969 R&S's parent company, Thompson-Starrett Company, merged with Elgin National Industries, Inc., taking the name of the latter corporation. Elgin National Industries' work is concentrated in two areas: 1) clocks and watches, and 2) specialized engineering, manufacturing and construction, which is performed by R&S and its subsidiaries.

R&S and two of its subsidiaries, J. O. Lively Construction Company and Langley & Morgan Corporation, design, engineer and construct plants. The remaining subsidiaries—Centrifugal & Mechanical Industries, Inc., Tabor Machine Company, Mining Controls, Inc., Clinch River Corporation, and Roberts & Schaefer Resource Service, Inc.—manufacture and sell preparation plant and mining equipment.

According to R&S, the company with its subsidiaries is able to provide the mining industry with anything needed to develop coal or mineral beneficiation plants and material-handling facilities. The company reports that it developed many of the technologies and systems that have evolved into today's sophisticated, automated preparation plants and coal-loading systems. For example, more than 20 years ago R&S secured exclusive license from Dutch State Mines to sell heavy-media cyclones in the United States. This equipment has enabled fine-size coal to be cleaned to a degree that was previously uneconomical. The company also states that it built some of the earliest high-tonnage unit-train loading equipment in the western coal-producing regions.

In 1979, 84 percent of Elgin National Industries' net sales and operating income came from work performed by R&S and its subsidiaries. The table that follows includes the revenues and operating profit of these companies.

Sales & Revenues (1980): $167,879,000

Operating Profit (1980): $15,339,000

Coal Cleaning Plants

Total number of plants constructed
or retrofitted (through 1980): over 2,000

       To prepare steam-grade coal: 75%
       To prepare metallurgical-grade coal: 25%

Number of contracts:

       1970 to 1975: 100
       1976 to 1980: NA

Plants profiled:

       Hamilton No. 1 Mine Preparation Plant,
       Island Creek Coal Company

Greenwich Collieries Preparation Plant,
Pennsylvania Mines Corporation/
Pennsylvania Power & Light Company

R & F Preparation Plant,
R & F Coal Company

Randolph Preparation Plant,
Peabody Coal Company

Manufactures preparation plant equipment

Coal Companies Owning
Preparation Plants

A.T. MASSEY COAL COMPANY
4 North 4th Street
Richmond, VA 23219

Coal Cleaning Plant Profiled: Marrowbone Development Corporation's Marrowbone Mine preparation plant, Naugatuck, West Virginia

Mines Served: Four coal mines Mingo County, West Virginia

Total Number of Cleaning Plants: 17

Coal Production and Percentage Cleaned: 14.1 million tons, 1980, (75% of this coal contained less than 1% sulfur; the balance had an average sulfur content of 1.5% or less), percentage cleaned NA.

Total Number of Mines: 69 (25 mining complexes) most of which are located in West Virginia, Kentucky and Tennessee.

Estimated Reserves: 900 million tons. Estimated that 90% of reserves contain less than 1% sulfur.

Customers: Almost all steam coal sold to domestic utilities. 80% of 1980 output sold under long-term contracts; of this, approximately 80% was sold as steam coal and 20% as metallurgical coal.

Financial Information - A.T. Massey (1980)

Sales Tonnage:

Coal Produced:	14.1 million
Coal Purchased:	8.3 million
Total:	22.4 million
Domestic Sales:	$439,491,000
Export Sales:	$287,511,000
Intersegment Sales:	$ 191,000
Net Coal Sales:	$722,192,000*
Net Income:	$ 22,216,000*

Parent Companies:	St. Joe Minerals Corporation and Scallop Coal Corporation
1980 Total Revenues: (St. Joe Minerals only)	$1,279,083,000
Coal:	$ 722,193,000
Lead, Zinc & Iron Ore:	$ 301,597,000
Oil and Gas:	$ 84,118,000
International Minerals:	$ 166,190,000
Others:	$ 4,985,000
Net Income:	$ 117,082,000
Scallop Coal Corp.	NA

*As of November 1, 1980, St. Joe has reflected one-half of Massey's sales, earnings and capital expenditures because of the new joint venture with Scallop Coal Corp.

A.T. Massey* has 17 preparation plants, all of which process coal with a sulfur content of less than 1.5 percent. The company's Marrowbone Mine preparation plant in Naugatuck, West Virginia— which is discussed in this profile— is the largest plant east of the Mississippi and one of the most innovative in INFORM's sample. This plant has two novel features: it is one of the earliest U.S. applications of filter press technology to dewatering wet fine-size particles, and it uses a flotation cell to clean fine-size coal particles (150 mesh to 0, the size of sand particles).

The use of filter presses eliminates the need for settling ponds. The

---

*A.T. Massey did not participate in INFORM's coal cleaning study. Win Holbrook, Vice President of Sales for A.T. Massey, said that INFORM's coal cleaning questionnaire required too much time from the company and that the information requested, if provided, might be taken out of context to be used against A.T. Massey by "environmentalists." Therefore, this profile is based on published information.

use of froth-flotation units (or flotation cells) to clean fine coal particles for utility use is a relatively new development. This equipment is different from most preparation plant equipment, which separates coal from its impurities by using differences in specific gravity. Rather, the Marrowbone equipment uses differences between the surface properties of coal and its impurities to effect their separation.

Massey became involved in coal preparation because the company believes that "the key to increasing sales and profits in the extremely competitive coal business is to match production to customers' specific needs and to avoid selling coal as just another commodity."

MARROWBONE MINE PREPARATION PLANT

In September 1979 Massey completed the Marrowbone Development Corporation mine complex in Mingo County, West Virginia. The Marrowbone complex includes four steam-coal mines, and production and preparaton of its low-sulfur coal is expected to reach 4 million tons per year in 1984, up from 1.2 million tons in 1980. Some of this coal is delivered to Carolina Power & Light Company (CP&L). (For more information on CP&L's use of cleaned coal, see CP&L profile.)

While A.T. Massey would not provide INFORM with the specific factors that induced the company to include filter presses in its plant, the reasons given by another company, Martin County Coal Company of Inez, Kentucky, suggest how important this equipment could be to some U.S. coal preparation plants. Martin County Coal had problems with sludge disposal because of the large amount of sludge (wet fine-size particles) to be disposed of, and because the rough topography of eastern Kentucky makes a sludge pond an expensive, short-term solution. Filter presses were chosen as a more permanent solution to the disposal problem. According to Raymond Bradbury, President of Martin County Coal, "The introduction of the filter press...will save us the money we would have spent finding and preparing additional spaces and impoundments to dispose of sludge."

A filter press takes wastewater carrying suspended fine particles, removes a clear effluent which can be recycled back to the plant, and produces a filter cake with a total moisture content that can be as low as 22 percent. These dewatered fine-size waste particles can then be disposed of with the coarse wastes from the preparation plant.

Filter presses are frequently used by the European coal industry to dispose of wastes in a limited area. In this country filter presses are used at municipal-waste and industrial facilities. They are now found at only three U.S. preparation plants (Marrowbone's, Martin County Coal's, and U.S. Steel's Cumberland Mine plant in Kirby, Pennsylvania).

Effectiveness of the Cleaning Plant

Coal Characteristics	Raw Coal	Cleaned Coal	Decrease / Increase
Sulfur (%):	NA	NA	--
Ash (%):	NA	NA	--
Moisture (%):	NA	NA	--
Btu per pound :	NA	NA	--

Btu recovery in cleaned coal: NA
Weight reduction in cleaned coal: NA

Note: The specifications set by Carolina Power & Light Company for coal delivered to its Roxboro Station (to which A.T. Massey supplies coal from its Marrowbone plant) are: 11 to 12 percent ash, a maximum of 6 percent moisture and 0.6 percent sulfur, and a minimum of 12,000 Btu per pound. (For more information on CP&L's Roxboro Station, see CP&L profile.)

Costs

A.T. Massey would not provide any cost figures for its Marrowbone coal cleaning operation.

MARROWBONE MINE PREPARATION PLANT

Beginning of Commercial Operation:	September 1979
Construction Time:	NA
Architectural, Engineering and Construction Firm:	Dravo Corporation
Total Tons Processed:	1.2 million tons produced in 1980
Customers:	Part of coal production to Carolina Power & Light Company

Manpower Required for Plant
Operation: NA

Efficiency of Plant Operations:

    Preventive Maintenance
    Requirements: NA

    Average Downtime: NA

Resources:

    Energy Rating: NA

    Water Use: NA

    Magnetite Requirements: NA

Wastes Generated: NA

ISLAND CREEK COAL COMPANY
P.O. Box 11430
Lexington, KY 40575

Coal Cleaning Plant Profiled:	Hamilton No. 1 Mine Preparation Plant Morganfield, Kentucky
Mine Served:	Hamilton No. 1 Morganfield, Kentucky

Total Number of Cleaning Plants: 28

Coal Production and Percentage
Cleaned: 20 million tons in 1980, 100% cleaned

Total Number of Mines: 26 (24 company-owned, 2 jointly-owned) located in Kentucky, Ohio, Pennsylvania, Virginia and West Virginia

Estimated Reserves: 3.8 billion recoverable tons; 1,559 million tons (45%) have a sulfur content of 1% or less

Customers:	Domestic	Foreign
Utilities:	11,099,000 (53%)	-------
Metallurgical Industry:	821,000 (4%)	5,340,000 (25%)
Industrial and Retail:	2,873,000 (14%)	813,000 (4%)
Total Tons:	14,793,000 (71%)	6,153,000 (29%)

Total Sales Tonnage:	20,946,000

Financial Information

Island Creek (1980):

Net Coal Sales:	$716,007,000
Total Revenues:	$720,920,000
Net Income:	$ 1,054,000

Parent Company:	Occidental Petroleum Corporation
Total Revenues (1980):	$12,726,312,000
Net Income (1980):	$ 710,785,000

Island Creek Coal Company, the nation's fourth largest coal producer in 1980, has had 50 years' experience in operating preparation plants. The company operates 28 plants, more than any other company in INFORM's sample, and cleans 100 percent of the 26 million tons of coal it mines. Island Creek's Hamilton No. 1 Mine preparation plant—discussed below—which includes heavy-media cyclones, represents a new trend in cleaning coal for utility use. As the price of coal rises and user specifications for coal quality become more strict, small-size coal particles can no longer be thrown away or added raw to the cleaned coal. The heavy-media cyclones used at Hamilton No. 1 effect a sharper separation between small-size coal (20 to 0 mesh) and waste particles than is possible with more traditional preparation plant equipment, such as jigs and heavy-media vessels.

Like Peabody, Island Creek, which sells more than half its coal to utilities, today has to respond to the needs of utilities that must meet increasingly stringent regulations governing sulfur dioxide and particulates, by providing a lower sulfur and ash, and a higher Btu, coal. This new customer demand for higher-quality coal is responsible for the fact that all but one of Island Creek's 28 plants are relatively sophisticated, treating coarse and fine particles to remove sulfur and ash, as well as mining wastes (cleaning to levels 3 and 4; see the chapter Methods of Coal Preparation). In 1979, to further improve the quality of coal sold to its long-term customers, Island Creek added circuits to treat fine coal particles in four of its existing plants in western Kentucky.

Design Expertise

Island Creek has gained so much coal cleaning expertise that it can play a major role in the planning and installation of its plants. Like three other companies in INFORM's sample, Island Creek does its own design work, developing a plant "flowsheet" and giving specifications for component size and for acceptable equipment manufacturers. "We'll listen to the suggestions of architectural/engineering/construction [A/E/C] firms," notes Elza Burch, Island Creek's Corporate Manager of Preparation, "but we've often had as much experience as they have. They don't really run the plants, we do; and as a result we can design a plant taking into consideration the problems of other plants." Island Creek uses A/E/C firms for plant-construction work.

Research and Development

Island Creek's preparation department engages in field research only, applying new equipment and processes to plants. "Cleaning very fine-sized coal particles (as small as they get) in heavy-media cyclones has been our most exciting development," states Burch. By adding centrifugal force to the heavy-media vessels, heavy-media cyclones clean small-size coal particles more effectively.

Island Creek also worked with Otisca Industries to develop the Otisca coal cleaning process which is now used by American Electric Power Company (AEP) to clean coal from its Muskingum Mine. Island Creek tested the Otisca Process in a 5 to 15 ton per hour pilot plant. (For a description of this process, see Otisca profile, and for further information on the Muskingum cleaning plant, see AEP profile.)

HAMILTON NO. 1 PREPARATION PLANT

Island Creek's original 1967 contract with Lively Manufacturing and Equipment Company (a subsidiary of Elgin National Industries), stipulated only a coarse-coal cleaning circuit. Island Creek just recently added the $6 million fine-coal cleaning circuit to Hamilton No. 1, and today the plant cleans coarse and fine-size coal particles.

The Hamilton No. 1 plant employs heavy-media cyclones to clean very small coal particles (as small as 20 to 0 mesh, or one twentieth of an inch), to produce a high-quality, more consistent coal, all of which is delivered to the Tennessee Valley Authority's Johnsonville generating station in Johnsonville, Indiana.

Heavy-media cyclones offer the flexibility needed to produce a uniform daily product from the fine-coal stream; the specific gravity of the separating medium can be changed to accommodate a wide range of feed-coal characteristics. According to Island Creek, this equipment gives Hamilton No. 1 the ability to meet customers' quality specifications with ease. (See chart on the operation of Hamilton No. 1 at the end of this profile.)

Island Creek would not give the capital cost of the coarse-coal circuit. The coarse-coal circuit treats 500 of the 800 tons fed through the plant each hour, and the fine-coal circuit cleans 300 tons per hour, or 37.5 percent of the feed, in a series of dense-media cyclones.

Effectiveness of the Cleaning Plant

Coal Characteristics	Raw Coal	Cleaned Coal	Decrease/ Increase
Sulfur (%):	3.85	2.80	-27.00
Ash (%):	21.84	8.90	-59.25
Moisture (%):	9.20	9.40	+2.00
Btu per pound:	NA	NA	NA

Btu recovery in cleaned coal: 96.5%
Weight reduction in cleaned coal: 32.4%

Waste Disposal and Cleaning Costs

Island Creek's Hamilton No. 1 plant produces a total of 3,500 tons of waste per day when the plant is operating at full capacity. The coarse refuse from Hamilton No. 1 is taken by truck to a dumping site where it is spread and compacted as Surface Mining Control and Reclamation Act regulations mandate. Island Creek would not provide information on the amount of land set aside for coarse-waste disposal or the length of time the company expects it to serve as a site. An impoundment has been built for fine refuse. Water from this impoundment is pumped back to the plant for reuse as the fine coal particles settle to the bottom of the pond. The costs of waste disposal were not available from Island Creek.

Although Island Creek would not give the annual operating and maintenance costs for the Hamilton No. 1 plant, the company gave figures—$1.00 to $1.50—for the cost of cleaning one ton of raw coal.

FACTORS INHIBITING AND ENCOURAGING THE DEMAND FOR CLEANED COAL

Island Creek believes that a lack of government action to support or enforce its avowed policy of promoting increased coal use by utilities is responsible for a lack of demand for coal, and hence for cleaned coal. As a result, the demand for new preparation plants is inhibited. In addition, congestion at shipping ports and the small number of such ports are noted as major impediments to greater U.S. coal exports. The lack of export facilities is of great concern to Island Creek, because the company has contracts for the sale of its cleaned coal to seven foreign countries—Britain, Romania, Italy, Spain, Japan, Egypt and Israel—and to parts of South America.

In an effort to alleviate its coal transportation and storage problems, another subsidiary of Occidental Petroleum has recently built coal terminals on the Ohio River, at South Shore, Kentucky, and Wheelersburg, Ohio. These facilties offer space and equipment for storage and blending and also offer access to domestic barge transportation, which is usually less expensive than rail transport.

According to Island Creek, there are several advantages to clean coal that encourage the growth of preparation plants. Prepared coal, Elza Burch claims, "may not be more expensive when it is considered part of the *entire* firing process clear through to waste handling, equipment design, and plant rating. The cost of transportation (which is absorbed by the customer) also argues against hauling 'reject' or unwanted material farther away from the production source than necessary."

FUTURE MARKET

The future of pre-combustion coal cleaning is assured, according to Island Creek. The market demand for good quality coal and innovations in preparation plant technologies, as well as environmental regulations, foster the growth of preparation plants. Coal cleaning will be needed not only as preparation for coal destined to be burned directly, but also for newer coal-transforming technologies.

"Whether we look ahead 5, 10, or 20 years, the clean air requirements will force pre-combustion coal preparation to some degree for all coals. Nobody will be able to put a completely raw product through any of the processes—direct combustion, chemical coal cleaning, liquefaction or gasification," maintains J. E. Katlic, Executive Vice President of Administration. Other Island Creek representatives assert that the plants of the future will be built for the intensive cleaning of coal. "To get sulfur out of the coal via coal cleaning before it goes into the boiler," Burch states, "whether or not it will also go through an FGD unit, makes the 'whole ball of wax' easier to handle."

In 1981, with the domestic coal market depressed, there is an excess of production capacity, but Katlic believes that as soon as the economy takes off again, so will the activity in coal preparation.

HAMILTON NO. 1 MINE PREPARATION PLANT

Beginning of commercial operation
    Coarse coal cleaning circuit: 1968
    Fine coal cleaning circuit: January 1979

Construction time
    Coarse coal cleaning plant: 1 1/2 years (18 months)
    Fine coal cleaning plant: Seven months

Architectural, engineering
and construction firm: . Lively Manufacturing & Equipment Company

Total tons processed: 800 tons cleaned coal per hour; approximately 2,538 million tons per year

Operates 13 1/2 hours per day; 235 days a year, or 4.7 days per week

Customers:

Manpower required for plant
operation: Six people per shift: three operators, one refuse-truck driver, two preparation plant engineers

Three lab technicians employed part-time

Efficiency of plant operations
    Preventive maintenance
    requirements: 40 hours per week with a three- to five-man shift

    Average downtime: 5% of total work time; attributed to equipment breakdowns

Resources

Energy rating:	9,000 horsepower
Water use:	10,000 gallons per minute, recycled through closed-loop circulation system; 250 to 300 gallons make-up water per minute
Magnetite requirements:	420 tons per month for dense-media cyclones

Wastes generated: 3,500 tons per day (30% of raw coal feed to plant): 50% from coarse-cleaning plant, 50% from fine-cleaning plant

COSTS:

Capital costs of plant

Coarse cleaning plant:	NA
Fine cleaning plant:	$6 million, 1979
Total:	NA

Annual operating and maintenance costs: NA

Costs of resources

Energy:	
Water:	Free from Ohio River
Magnetite:	$0.05 per pound; $42,000 per year: 1979

Waste disposal costs: NA

Price charged per ton
of cleaned coal: NA

Transportation cost per ton
of cleaned coal: NA

Financing source for plant: Funding for all of Island Creek's preparation plants comes from internal cash flow or from the parent company, Occidental Petroleum Corporation. Plants are generally amortized over 20 years. Investment tax credits are given by government for plant construction.

MONTEREY COAL COMPANY
P.O. Box 496
Carlinville, IL 62626

Coal Cleaning Plant Profiled:	Monterey No. 2 Preparation Plant Albers, Illinois
Mine Served:	Monterey No. 2 Mine Albers, Illinois

Total Number of Cleaning Plants: Three

Total Number of Mines: Three, two located in Illinois and one located in West Virginia

Coal Production:

Monterey No. 1 Mine (1980):	2,913,264 tons
Monterey No. 2 Mine (1980):	2,089,261 tons
Wayne Mine (1980):	42,000 tons (initial production)
Total Tons (1980):	5,044,525 tons

Production Capacity:

Monterey No. 1 Mine:	3 million tons per year
Monterey No. 2 Mine:	3 million tons per year
Wayne Mine:	2 million tons per year

Parent Company Recoverable
Reserves: 9,500 million tons in Illinois, West Virginia and Wyoming

Customers:

Steam Coal:	100%
Metallurgical Coal:	0%

Financial Information - Monterey Coal Company

Parent Company: Exxon Corporation

Total Revenues:	$110,380,629,000
Oil, Gas, Coal and Chemical Products:	$108,448,834,000
Other (interest):	$ 1,931,795,000
Net Income:	$ 5,650,090,000

Monterey Coal Company, a subsidiary of Exxon Corporation, owns two coal preparation plants in Illinois and one in West Virginia. The Monterey No. 2 Preparation Plant, located in Albers, Illinois, cleans its coal to a level (level 3) that is achieved by slightly over a quarter of the plants in the United States.* A typical midwestern plant, it cleans raw coal with wash boxes (jigs), removing about 15 percent of the sulfur and 68 percent of the ash, and uses screens and centrifuges for water removal. The plant's automatic sampling equipment is described as "state of the art" by the company.

MONTEREY NO. 2 PREPARATION PLANT

Monterey No. 2, which began operating in 1977, was designed, engineered and built by McNally Pittsburg Manufacturing Corporation. The plant processes 100 percent of the coal from the Monterey No. 2 Mine. It cleans coal that is less than two inches in size, but it does not use a fine-coal circuit (particles smaller than 3/16 inch are shipped uncleaned with the larger, cleaned and dewatered coal sizes). The cleaned coal, crushed to 1½ inch to 0 before shipment, is mixed with the fine coal and loaded into two 7,500-ton unit trains which haul it to the Gibson Generating Station of Public Service Indiana (see Public Service Indiana profile), 117 miles away.

No coal is stored in open piles; a source of wind-blown dust is thus eliminated.

Water Use and Waste Disposal

Water used in the Monterey plant is recycled. Waste slurries from the plant are allowed to settle in a receiving pond nearby; water is then clarified in an on-site treatment plant and stored in a 30-acre, 65,000,000-gallon recirculation pond until reuse. About 2,000 gallons of water per minute are added to the cleaning circuits from a lake located on the plant property, to make up water losses from evaporation and absorption.

---

*This figure was determined by R. Schmidt in a report, "Coal Preparation Plants," done for the Electric Power Research Institute.

According to Monterey, waste materials comprise 30 to 35 percent of the plant's raw coal feed, and an average of 6,172 tons of waste is produced daily. Solid wastes are moved by conveyor to a 400-ton waste bin. From there the wastes are transported by truck to a nearby disposal site where they are dumped, spread and compacted. Runoff water is collected from this site and returned to the clarification circuit for treatment.

Effectiveness of the Cleaning Plant

The following table lists the sulfur, ash, moisture and Btu content of Monterey's raw and cleaned coal, as well as the percentage of Btu's originally present in the raw coal that are recovered in the cleaning process:

Coal Characteristics	Raw Coal	Cleaned Coal	Decrease / Increase
Sulfur (%):	3.9	3.3	-15.4
Ash (%):	25.0	8.0	-68.0
Moisture (%):	13.5	15.5	+14.8
Btu per pound :	8,400	10,800	+28.6%

Btu recovery in cleaned coal: about 90%

Costs

The capital cost of the Monterey No. 2 preparation plant was $25 million in 1977 dollars. This figure includes the cost of the plant's front- and back-end loading facilities. No information was available on the operating and maintenance, resource, and waste disposal costs, or on the price of the plant's cleaned coal.

FUTURE MARKET

Monterey Coal Company cleans its coal in order to improve its quality and thus "maintain flexibility to sell to as many customers as possible." Because two of Monterey's mines draw on the high-sulfur and -ash coal of the Illinois Basin, the quality of the company's raw coal is not as high as that offered by many eastern and western coal suppliers.

MONTEREY NO. 2 PREPARATION PLANT

Beginning of Commercial Operation: July 1977

Construction Time: Approximately three years

Architectural, Engineering and
Construction Firm: McNally Pittsburg Manufacturing
                                          Corporation

Total Tons Processed: Twin 1,000 ton per hour circuits
                                          for raw coal; clean coal production:
                                          1,350 tons per hour; 1,297,744 tons
                                          produced in 1979, 2,089,261 tons
                                          produced in 1980

Customers: Public Service Company of Indiana,
                                          Inc.; Gibson Generating Station

Manpower Required for Plant
Operation: Nine men per shift; two cleaning
                                          shifts and one maintenance shift per
                                          day

Efficiency of Plant Operations

   Preventive maintenance requirements: Eight hours per day

   Average downtime: NA

Resources

   Energy rating: 7,500 horsepower

   Water use: 2,000 gallons per minute make-up
                                          water

   Magnetite requirements: None

Wastes Generated: 30% to 35% reject rate; average of
                                          6,172 tons per day

COSTS:

Capital Cost of Plant and Coal-
Handling Facilities: $25,000,000

Annual Operating and Maintenance Costs: *

Costs of Resources

   Energy: *

   Water: *

   Magnetite: *

Waste Disposal Costs: *

Price Charged Per Ton of
Cleaned Coal: *

Transportation Costs Per
Ton of Coal: *

Financing Source for Plant: *

*Confidential information

NORTHWEST COAL CORPORATION
315 East 200 South
Salt Lake City, UT 84111

Coal Cleaning Plant Profiled: Western Slope Carbon Preparation
                                 Plant
                                 Somerset, Colorado

Mine Served: Hawk's Nest Mine
                                 Somerset, Colorado

Total Number of Cleaning Plants: One

Coal Production and Percentage
Cleaned (1980):   435,000 tons, 100% cleaned

Total Number of Mines: One (five seams situated over
                                    four federal leases)

Estimated Reserves: 40 million tons

Customers: Central Illinois Light Company;
                                    CRA Corporation

Financial Information - Northwest Coal (1980)

    Sales Tonnage: 863,000
    Net Coal Sales: $23,600,000
    Net Income (Loss):   ($3,800,000)

Parent Company: Northwest Energy Company
                                    (principal asset: Northwest Pipeline
                                    Corporation, an interstate natural
                                    gas transmission company)

Total Revenues (1980): $1,323,432,000
Net Income (1980): $ 54,075,000

Northwest Coal Corporation is a relatively new entrant in the coal field. Formed as a wholly owned subsidiary of Northwest Energy Company in 1974 to carry out Northwest Energy's long-term plans for coal production, Northwest Coal purchased Western Slope Carbon, Inc. and its one operating mine, Hawk's Nest, in Somerset, Colorado, for $12 million in the same year. While seeking long-term contracts for the sale of its coal, Northwest Coal discovered that although the fuel had a high Btu and very low sulfur content (12,000 Btu per pound and 0.6 pounds of sulfur per million Btu), its high ash content eliminated many utilities and industries as potential buyers. To insure the success of its first coal venture, Northwest Coal's management decided to improve the coal by cleaning it.

In 1979, to fulfill its anticipated contract with Central Illinois Light Company (see CILCO profile), Northwest Coal contracted with the architectural/engineering/construction firm of Allen & Garcia for a preparation plant. Designed to enable Northwest Coal to meet the ash specification set by CILCO, the Western Slope Carbon Preparation Plant reduces the ash in its coal from 20 to 8 percent.

Two constraints were placed on the design of the plant. First, it had to be built quickly so that Northwest could meet CILCO's ash requirements by January 1980 (the contract for delivery of cleaned coal to CILCO was actually signed on October 30, 1979), and second, it could occupy very little level land, as the Hawk's Nest Mine is located in a narrow canyon. Hence, the Western Slope Carbon Plant, built in 1979, was one of the most quickly constructed preparation plants in history (excluding the modular, portable plants). The space limitations resulted in a unique plan for underground waste disposal. However, this plan was later discarded (see the Waste Disposal section of this profile).

When the preparation plant was built, Allen & Garcia also built a unit train load-out facility. Its twin towers can hold 8,000 tons of coal and load the 73-car unit trains in less than two hours. According to Northwest Coal, this facility enables it to offer customers lower transportation rates.

Today Northwest Coal has two contracts for the sale of its cleaned coal: one with CILCO to deliver 600,000 tons per year for the next 15 years, and the second to deliver from 30,000 to 100,000 tons per year to the CRA Corporation (a subsidiary of Farmland Industries), which uses the coal to fire boilers for the refining of crude oil. Deliveries to CRA began in January 1981.

Western Slope Carbon Preparation Plant

The Western Slope Carbon Preparation Plant was completed only 10 months after the start of construction. The design, engineering and construction work for a preparation plant usually takes from one to three years. However, as noted above, Northwest Coal was pressed for time because of contractual obligations, and work was begun on the plant in early 1979, before all the environmental permits for its operation and disposal of its wastes had been obtained. This was an unusual sequence. Peter Matthies, Vice President of Northwest Energy Company, stressed that a lot of attention was given to complying with every detail of the law, to minimize the risk inherent in building a plant before getting permits to operate it. The outstanding permits for plant operation and waste disposal were received from the U.S. Office of Surface Mining and Reclamation and the Colorado Mine and Reclamation Board 6 to 12 months after construction of the plant began.

Effectiveness of the Cleaning Plant

The following table lists the characteristics of raw and cleaned coal handled by the Western Slope Carbon Preparation Plant:

Coal Characteristics	Raw Coal	Cleaned Coal	Decrease / Increase
Sulfur (%) :	0.6	0.6	0
Ash (%) :	20.0	8.0	-60
Moisture (%) :	5.0	7.5	+50
Btu per pound :	12,000	12,600	+ 5 %

Btu recovery in cleaned coal: 90%
Weight reduction in cleaned coal: 20% to 25%

The plant, which uses a simple Baum jig for washing, and centrifuges for dewatering 5 inch to 0 coal particles, is capable of making three products: "slack," "stoker" and "lump" coal (for definitions of these terms, see Appendix).

Waste Disposal

INFORM profiled the Western Slope Carbon Preparation Plant partly because Northwest Coal had developed a novel plan—which it has since decided not to implement—to develop an underground waste disposal system in mined-out portions of the Hawk's Nest Mine with the help of funds from the Bureau of Mines. No other company profiled by INFORM has an underground disposal system. Tailings from the plant were to be transported via slurry pipeline to the mine at a volume of up to 1,216 tons per day. There the water would have been separated from the solid refuse by cyclones, and returned to the plant for use as make-up water. The solids would be blown pneumatically into the underground disposal area. This system was to solve the waste disposal problems arising from the limited space available by utilizing three processes which have never been used in combination in the United States: slurrying, dewatering and pneumatic blowing.

The company decided not to go through with this plan because it believed that waste handling at the mine site would require more men and machinery, causing confusion and congestion, and a loss of productivity at the Hawk's Nest Mine. Northwest Coal would also have to make sure that it had mined enough to make room for the wastes.

Only the permitting process had been started when Northwest Coal abandoned this plan. The company estimated the cost of underground disposal for a ton of refuse, but it considers the figures unreliable and would not release them. The estimated capital cost for this system was $2 million.

At present Northwest Coal is trucking its wastes from the preparation plant to a county landfill 26 miles away. According to Al Amundsen, Chief Engineer of Northwest Coal, the cost of waste disposal is high, $3.00 a ton, because the refuse has to be moved so far; the company is now seeking land for a waste pile closer to its plant. Northwest Coal is responsible for handling the wastes safely, insuring that there is no leaching or dust emissions from the landfill.

Costs

The capital cost of Northwest Coal's preparation plant was $2.2 million in 1979. Figures for the annual operating and maintenance costs, and for the price charged per ton of cleaned coal, were not available from the company.

According to Matthies, the cost of the plant is partially offset by the increased productivity of the Hawk's Nest mine. Before the preparation plant was completed, Matthies stated, Northwest Coal had to mine selectively, so coal would be brought up with fewer wastes and CILCO would be assured of a fuel with no more than 12 percent ash. Now that the plant is operating, coal cleaning alone reduces the coal's ash content. Eliminating the need for selective mining brings up a coal averaging 20 percent ash, but also increases coal production per man-hour.

Factors Inhibiting and Encouraging the Demand for Cleaned Coal

Amundsen cites three factors that inhibit the increased use of preparation plants: lack of demand for coal, environmental regulations and strikes. The worst culprit in the situation, according to Matthies, is the government, which has extended regulations to cover every detail of mining and preparation plant activities.

In addition to improving its quality, another incentive for cleaning coal, claims Matthies, is the reduction of customers' transportation costs. When the weight of the coal is diminished by 20 to 25 percent through washing, costs for transporting coal 1,350 miles to Central Illinois are correspondingly lowered.

Future Market

Amundsen believes that market conditions which demand high quality coal and environmental regulations governing utility and industrial emissions, encourage the use of preparation plants. He says that future plants will be built to clean coarse, medium and fine-size coal particles to meet these demands.

In the future, Northwest Coal will not open a mine without a preparation plant, according to Al Perry, the corporation's marketing manager. Northwest Coal sees utilities and the western "industrial stoker'' market as its future customers. If it is to capture some of the more lucrative industrial stoker business, it must clean its coal to provide a very high quality product (approximately 13,000 Btu, 5 percent moisture, 5 percent ash and 0.65 percent sulfur). By improving the Btu content of the fuel, cleaning low-sulfur coal from the Hawk's Nest Mine gives Northwest an edge over its competitors selling to utilities as well.

Northwest Coal also sees prospects for exporting the coal from its recently acquired Utah reserves. Matthies notes that potential buyers in Japan and Korea have expressed an interest in Northwest's coal. ''While a contract for coal with Korea might be a ways off,'' he stated in November 1980, ''I am heading to Japan [to consult a potential customer] tomorrow p.m.''

WESTERN SLOPE PREPARATION PLANT

Beginning of Commercial Operation:	January 1980
Construction Time:	Eight months
Architectural, Engineering and Construction Firm:	Allen & Garcia Company
Total Tons Processed:	400 tons per hour; 4.1 million tons per year
	Operates 14 hours per day, five days per week
Customers:	Central Illinois Light Company and CRA Corporation
Manpower Required for Plant Operation:	960 man-hours per month to operate the plant
	320 man-hours per month to maintain the plant
	320 man-hours per month to supervise operations
Efficiency of Plant Operations	
Preventive maintenance requirements:	320 hours per month
Average downtime:	From January to July 1980, 10% of operating time; attributed to problems in the refuse-handling system

Also, oversized rocks (8 inches and larger) frequently jammed the system. In July 1980, a new crusher was installed before the refuse circuit, eliminating this problem

Resources

Energy rating:

550 horsepower, 87 gallons of make-up water per minute

Water use:

13 gallons per ton of coal. The bulk of this water is recycled through a closed-loop circulation system

Magnetite requirements:

None. No heavy-media cleaning equipment in plant

Wastes Generated:

1,216 tons per day (20% to 25% of raw coal feed to plant); 85% of the refuse is coarse and 15% fine

COSTS:

Capital Cost of Plant:

$2.2 million, 1979

Annual Operating and Maintenance Costs:

NA

Costs of Resources

Energy:

NA

Water:

NA

Magnetite:

NA

Waste Disposal Costs:

NA

Price Charged Per Ton of Cleaned Coal:

NA

Transportation Costs Per Ton of Coal:

NA

Financing Source for Plant:

Funding for the preparation plant came from the parent company, Northwest Energy Company. Investment tax credits given by government for plant construction

PEABODY COAL COMPANY
301 North Memorial Drive
St. Louis, MO 63102

Coal Cleaning Plant Profiled:	Randolph Preparation Plant Marissa, Illinois
Mine Served:	Baldwin No. 1 Marissa River King Pit Marissa, Illinois

Total Number of Cleaning Plants:	18
Coal Production and Percentage Cleaned:	60.2 million tons in 1979 (almost 8% of the nation's total production), 50% cleaned
Total Number of Mines:	46 mines in ten states: 28 surface mines accounting for 2/3 of company production; 18 underground mines accounting for 1/3 of company production
Estimated Reserves:	9 billion tons

Customers (1980):

Utilities:	90%
Industrial:	10%

Financial Information - Peabody Coal (1980)

Coal Sold, Short Tons:	55,346,000
Coal Sales Revenues:	$1,134,805,000
Net Income:	$ 90,090,000
Parent Company:	Peabody Holding Company, Inc. (PHC), a consortium of six companies including: Newmont Mining Corporation (27.5%), Williams Companies (27.5%), Bechtel Corporation (15%), Boeing Corporation (15%), Fluor Corporation (10%), and Equitable Life Assurance Society (5%)
Parent Company Total Revenues (1980):	NA
Parent Company Net Income (1980):	NA

Peabody Coal Company,* the largest coal producer in the country, owns and operates 18 preparation plants in four states. (For the tonnage processed at each of these plants and their locations, see table and map on preceding page.) In 1979 Peabody cleaned approximately half its total production of 60.2 million tons. One plant, the Randolph Preparation Plant in Marissa, Illinois—which is described below—is an example of a cleaning facility (a jig-hydrocyclone plant) which was built to treat coarse, medium-size and fine-size particles produced from coal mined underground. This plant treats the largest quantity of raw coal of any single preparation plant in the United States (2,200 tons per hour).

More than 90 percent of Peabody's output is sold to utilities. The company must respond to their demands for cleaned coal. As a result, a company official noted, in the future "far more than half of Peabody's coal will have to be cleaned," and much of it will be cleaned with more sophisticated (heavy-media) equipment.

Considerations in Preparation Plant Design

Peabody, like the other older and larger coal companies profiled by INFORM (see Island Creek and Pittston) does much of its own design work. When a decision is made to build a preparation plant, Peabody's coal cleaning staff generally develops a "flowsheet" and designates sizes for the equipment to go into the plant. Peabody then takes this flowsheet to a preparation plant builder to get a bid on the construction costs. Peabody often specifies three name brands that it regards as acceptable for a particular piece of equipment. According to a Peabody official, offering some choices to the plant constructor allows for competitive bidding.

Peabody is reported to have recently developed a detailed record-keeping system to improve preparation plant design. A series of production and maintenance records for each plant is being kept to pinpoint lost operating time and its causes. The lost operating time is calculated in minutes and hours, and also in terms of the coal which could have been cleaned at the plant had it been operating. This information indicates

---

*Peabody did not cooperate with INFORM on this coal cleaning study. Information in this profile is based on a published article in *Coal Age*, publicly available company documents, and phone conversations with a company official.

the operating cost of a particular piece of equipment and will be useful whenever Peabody plans another preparation plant.

Meeting Stricter Specifications

Until recently, the primary purpose of coal cleaning was to reduce ash and gross impurities. The most important consideration in designing a preparation plant was the size and variation in size of the raw coal particles to be treated. To reduce ash as well as mining wastes, jigs were most effective in screening the larger-size surface-mined coal, while jig-hydrocyclones were better at cleaning the smaller-size coal particles produced by a "continuous" underground mining machine. Seventeen of Peabody's plants use jig or jig-hydrocyclone cleaning equipment, and each type processed half the 30.1 million tons of raw coal the company cleaned in 1979. (For a description of jig and jig-hydrocyclone plants, see the chapter Methods of Coal Preparation.)

Today, however, Peabody says that government regulation of utility emissions and the increasing cost of preparation plants (due to inflation) are two new factors that must be considered in designing, engineering and constructing preparation plants. Stricter coal specifications set by utilities as a result of increasingly stringent sulfur dioxide emission standards are, according to Peabody, as important a consideration in preparation plant design as the size of the coal pieces.

Camp 9, Peabody's newest and first heavy-media preparation plant, was designed in response to the challenge of these stricter coal specifications. Located in Waverly, Kentucky, Camp 9 (TVA-owned, Peabody operated) was opened in October of 1980. Heavy-media cleaning equipment was chosen because when finely ground magnetite is added to the water used to separate coal and waste particles, this equipment can accomplish a more exact separation between them than jigs alone or in combination with hydrocyclones. Heavy-media equipment can potentially remove more of the ash and sulfur in coal. George Morris, Director of Coal Preparation for Peabody, said, "I favor heavy-media for coarse cleaning rather than a jig, and hydrocyclones for the fine coal. It gives us flexibility and permits us to change the gravity for the plus ¾ " coal on an hourly basis. When used with hydrocyclones, we have a lot of flexibility to produce a uniform daily product."

Other advantages often claimed for the heavy-media cyclone with the hydrocyclone are that a minimal screen area is required and less magnetite is consumed.

As a partial answer to the problem of rising costs of preparation plants due to inflation, Peabody is designing, engineering and constructing dual-circuit preparation plants. Dual-circuit plants are simply two identical preparation plants built side by side. Peabody claims that these plants give the company more flexibility than a single-circuit plant. The two circuits can be built at the same time, or only one can be built, along with the parts that would be used by both circuits, such as the conveyor system. This enables management to postpone some capital costs if the cost of borrowing money is high. Until the mine serving the preparation plant reaches full production and both circuits are needed to operate full-time, one circuit can run for two shifts, or two circuits for one shift each.

RANDOLPH PREPARATION PLANT

Peabody's Randolph Preparation Plant in Marissa, Illinois—engineered, built and later expanded by the architectural, engineering and construction firm of Roberts & Schaefer—delivers cleaned coal to the Associated Electric Cooperative's New Madrid generating station in Marston, Missouri. Peabody claims that a recent modernization effort at the plant has improved coal quality by 300 to 400 Btu per pound.

Effectiveness of the Cleaning Plant			
Coal Characteristics	Raw Coal	Cleaned Coal	Decrease/ Increase
Sulfur (%):	NA	NA	--
Ash (%):	NA	NA	--
Moisture (%):	NA	NA	--
Btu per pound:	NA	NA	--

Btu recovery in cleaned coal: NA
Weight reduction in cleaned coal: 26.8%

The Randolph plant now processes Illinois seam No. 6 coal from Baldwin No. 1 and Marissa underground mines and the River King Pit surface mine. The 3,000 tons of raw coal that the Marissa mine now produces per day is processed into approximately 1,800 tons of clean coal. Marissa is the first of three mines to open, which together are known as the TSM complex (Tilden, St. Libory, Marissa). At full production, the

TSM complex is expected to produce 3.5 million tons of coal per year for the next 25 to 30 years, all of which is to be cleaned at the Randolph plant. A reclaim feeder will move coal from the mine complex (from a 30,000-ton raw-coal storage "funnel") to the six-mile overland conveyor belt that carries it to the Randolph plant.

Waste Disposal and Cleaning Costs

While Peabody did not officially provide information on its provisions for disposing of the 8,834 tons of waste produced each day from its Randolph plant, or on the costs of building or operating the plant, one manager gave ball-park figures for the additional costs per ton of a cleaned coal over a raw coal: about $10 per ton more for a deep-cleaned coal and $5 per ton more for a minimally cleaned coal.

RANDOLPH PREPARATION PLANT

Beginning of Commercial Operation:	NA
Construction Time:	NA
Architectural, Engineering and Construction Firm:	NA
Total Tons Processed:	2,200 tons of raw coal per hour; 1,676 tons cleaned coal per hour
Customers:	NA
Manpower Required for Plant Operation:	Two eight-hour operating shifts per day
Efficiency of Plant Operations	
Preventive Maintenance Requirements:	One eight-hour maintenance shift per day
Average Downtime:	NA
Resources	
Energy Rating:	NA
Water Use:	Eight gallons of recycled water per ton of coal

Magnetite Requirements: NA

Wastes Generated: 8,384 tons per day

THE PITTSTON COMPANY
One Pickwick Plaza
Greenwich, CT 06830

Coal Cleaning Plant Profiled: Rum Creek Preparation Plant
                                 Logan County, West Virginia

Mines Served: Six

Total Number of Cleaning Plants: 23

Coal Production and Percent   5,664,000 tons steam coal (32%)
Cleaned: 12,112,000 tons metallurgical coal
                                              (68%)
                                  17,776,000 net tons produced, 80%
                                              cleaned

Total Number of Mines: 59

Estimated Reserves: 2.9 billion tons of recoverable re-
                                  sources in Virginia, West Virginia
                                  and eastern Kentucky

                                  1.6 billion tons of "proved reserves"
                                  (or coal that can be profitably ex-
                                  tracted using present mining tech-
                                  niques and equipment)

                                  Of the above 1.6 billion tons:
                                  62% has a sulfur content of less
                                  than 1%; 27% has a sulfur content
                                  of between 1% and 1.5%

                                  400 million tons in Wyoming

Customers:
                                  Domestic   Export
    Utilities and Industrial: 6,554,000 tons   429,000 tons
    Metallurgical Industry: 920,000 tons   10,839,000 tons
    Total: 7,474,000 tons   11,268,000 tons

Total Sales: 18,742,000 tons

Financial Information (1980):

Pittston Company and Subsidiaries:

Net Sales & Operating Revenue:	$2,051,611,000
Net Income:	$ 75,802,000

Revenues (1980):

Coal:	38%
Oil(fuel oil distribution):	52%
Brink's(worldwide security transportation)	10%

The Pittston Company ranked sixth among all U.S. coal producers in 1980, producing 17,776,000 tons of coal, representing two percent of the country's production. Pittston's reserves are rich in clean-burning coal (62 percent of Pittston's "proved reserves" have less than one percent sulfur) from underground mines in the Appalachian Mountains. Eighty percent of Pittston's coal was cleaned to levels three, four and five in its 23 preparation plants before sale. (Pittston also owns and operates an additional 12 plants that only crush and size coal, which the company calls "docks," "tipples" or "loading sites," rather than preparation plants.)

The Rum Creek plant in West Virginia is Pittston's newest preparation plant, incorporating many innovations. It is efficient, is especially flexible (it can process coal of very different qualities and produce two end products), and conforms to both present and anticipated environmental regulations. It includes new equipment such as froth-flotation units and is one of the most advanced, automated plants in INFORM's sample.

Marketing Strategy

Pittston's 23 preparation plants represent the largest coal preparation plant network in the Appalachian region and are an integral part of the company's coal marketing strategy. To make coke, the steel industry demands a high-quality, uniform "metallurgical" coal; also, cleaning "steam" coal generally provides utility and industrial buyers with a lower-ash and better quality coal which commands a higher price. Pittston originally became involved in coal preparation in the late 1940s so that it could sell its coal to the steel industry. Today Pittston sells approximately 40 percent of its coal to utility and industrial users, and about half of this is cleaned.

Pittston has suffered from the downturn in world steel markets and overcapacity in the industry. Nicholas T. Camicia, President and Chairman of Pittston, hopes to increase the market for the company's coal by selling more of its low-sulfur metallurgical coal to electric utilities.

The design of the Rum Creek plant reflects Pittston's expectation of selling more steam coal in the 1980s. Rum Creek was one of four new plants brought into operation in 1979 and 1980 that were based on the same flexible design. The other preparation plants were Beckley No. 2 and Grand Badger in West Virginia and McClure No. 1 in Virginia. These new plants are designed to do all kinds of blending, so that there can be a wide range of finished-product possibilities (from a metallurgical coal, to a premium steam coal, to a lower-quality steam coal).

Recently Pittston got its steam-coal business rolling by signing contracts for "compliance" coal that is so low in sulfur that it can comply with utility sulfur dioxide requirements under the federal New Source Performance Standards (NSPS) without the need for scrubbers (flue gas desulfurization systems). Compliance coal must be cleaned so that the sulfur content of coal shipped to a client does not exceed a specific percentage. In 1978 Pittston signed a contract with the Tennessee Valley Authority for compliance coal, and in 1980 the company signed a similar contract with Appalachian Power Company, a subsidiary of AEP, for coal from the Rum Creek Preparation Plant.

Pittston is the leader in the U.S. coal export market. In 1980 it shipped 11,268,000 tons of mostly metallurgical coal to customers in Japan, Europe, Canada, Algeria and South America. The company's share represented 13 percent of the U.S. coal export market, which reached a record total of over 90 million tons in 1980.

Although there was a surge in U.S. exports of steam coal to Europe in 1980, Pittston said that it did not participate on a large scale because coal buyers were mostly seeking spot-market purchases of low-quality and low-priced coal without long-term commitments. However, beginning in 1981 Pittston did make its first step into the steam-coal export market, with contracts to ship more than 1 million tons of coal a year to Europe. Domestically, Pittston's 1980 sales of steam coal were 13 percent over the 1979 figure.

Considerations In Preparation Plant Design

Pittston's preparation plants are designed in a joint effort by its Coal Group's preparation plant department and the architectural, engineering and construction (A/E/C) firm selected to engineer and construct the plant. Pittston chooses an A/E/C firm on the basis of past performance, cost, and review of the firm's engineering proposals for the plant.

Research and Development

Pittston engages in applied research rather than developmental laboratory research. Pittston was the single supplier of coal to Florida Power & Light Company for that company's research on blending coal with oil to create a coal-oil mix (COM) for use in oil-fired utility and industrial boilers. (A COM mixture contains approximately 50 percent coal and 50 percent oil). This was one of the largest COM experiments in the United States.

Although Florida Power & Light Company paid for Pittston's coal, Pittston claims that it aids the utility by providing chemical- and physical-property analyses of this coal. In addition, Pittston is providing coal for COM tests at several major energy labs. If COM could be made commercially successful, it could reduce the amount of oil needed in generating power by 50 percent in each boiler that uses it. Pittston sees it as another potential market for its low-sulfur cleaned coal.

Pittston has also participated in several other research projects, supplying free coal samples and developing technical data on the coal. One such sample was given to an Atlantic Richfield Company-Electric Power Research Institute desulfurization project. Pittston maintains that it provides free coal samples whenever it sees the processes being developed as opening potential markets for its own coal. If the free samples are large, the companies receiving them generally pay transportation costs.

Some of Pittston's research and development efforts to improve conventional coal preparation include working with suppliers of preparation plant equipment. Pittston will often field-test a piece of equipment for a supplier at one of its plants, where an on-site analysis of the effectiveness of the equipment in plant performance can be done. Pittston is also working with the Bureau of Mines to develop an electrostatic coal separator which could be used to reclaim the Btu's in very fine coal particles from ''blackwater'' (the waste-disposal ponds at preparation plants).

RUM CREEK PREPARATION PLANT

The Rum Creek Preparation Plant in Logan County, West Virginia, is operated by Pittston's subsidiary, the Elkay Mining Company, and processes raw coal from mines in the Upper Cedar Grove, Stockton, Coalburg, Eagle, and 5-Block seams. Of the 1.6 million tons of cleaned coal produced annually at Rum Creek, 1 million tons a year have been promised for the next 20 years to Appalachian Power Company (APCO) for approximately $41 per ton under a 1980 contract. The contract contains an escalation clause for the price of coal; according to Pittston, the terms of the clause are confidential.

Pittston has guaranteed its customers that when the coal from Rum Creek is burned, its sulfur dioxide emissions will not exceed 1.2 pounds per million Btu. This enables APCO's new Mountaineer plant in New Haven, West Virginia, to meet the NSPS without scrubbers.

Effectiveness of the Cleaning Plant

Coal Characteristics	Raw Coal	Cleaned Coal	Decrease/ Increase
Sulfur (%):	NA	NA	--
Ash (%):	NA	NA	--
Moisture (%):	NA	NA	--
Btu per pound:	NA	NA	--

Btu recovery in cleaned coal: 80%
Weight reduction in cleaned coal: 40%

The Rum Creek plant washes both large and small coal fractions. It uses heavy-media vessels, heavy-media cyclones, froth flotation and hydrocyclones to process 826 tons of raw coal per hour, to produce high-quality steam and metallurgical coals.

Innovative Features of Rum Creek

According to Pittston, the Rum Creek plant has several special features. One characteristic that makes the plant efficient and flexible is a large medium-cleaning circuit ("middling" circuit). "We put a middling circuit in our plants so we can pick a specific gravity that won't throw any coal away. This lets us make a metallurgical-coal product and a steam-coal product, or a premium steam-coal fraction and a medium fraction,'' says a Pittson representative. The products can be sold separately or blended in any proportion portion to meet various needs. At Rum Creek, the cleaned 1½ inch to 0 metallurgical and premium steam coals go to separate 5,000-ton-capacity silos. By varying the feed rate from each silo and storage area and closely monitoring a belt scale and a Sortex Ash Monitor located at the load-out, the operator can tailor coals to almost any customer specification.

Another special feature of Rum Creek addresses the problem of processing coal from many sources. Large variations in the quality of the coal fed to this plant could potentially impair the performance of the froth-flotation units. To minimize the loss of coal from improper froth-flotation performance, hydrocyclones are included in the plant circuit ahead of these units to insure a more consistent-size feed.

The Rum Creek plant is designed to comply with all environmental and safety restrictions, present and anticipated, as are the other three new Pittston preparation plants. To insure compliance, a company spokesman notes, the Rum Creek plant contains the latest equipment for protection against every hazard, from excessive noise to pollution from dripping conveyor belts. Dry-dust collectors, sometimes called baghouses, reduce airborne dust, and rubber-lined chutes and screens reduce noise. The thermal dryer system, which is used to keep the moisture of the final product consistently low, employs equipment to prevent excessive stack emissions. This equipment, similar to that found in electric power plants, involves a precipitator that removes 92 to 95 percent of the larger particles, followed by a wet scrubbing system that guarantees an emission rate of less than 0.03 grain per cubic foot of particulates.

The automation and instrumentation of the Rum Creek cleaning circuit also reflect how advanced this plant is. Automation and instrumentation present many benefits not available to plants without such sophisticated design. They offer more efficient use of manpower and enable the operator to control more accurately the quality of the coal end product. With little monitoring by the plant operator, an automatic heavy-media circuit at Rum Creek maintains the specific gravity used to separate waste and coal particles at within 0.02 of that desired. Visual and electronic monitors are available to the plant operator to check on any of the plant's operations. These monitors allow an operator to identify and correct problems more quickly, improving plant performance and efficiency.

Controls that are capable of varying the rate of coal feed to the plant, and scales that continuously monitor and determine the total raw coal feed and clean coal product, enable the operator to adjust the load of feed to the plant in order to suit specific operating conditions and to offer greater system flexibility for desired changes in the end product. Continuous monitoring of both the feed to the thickener and of the clarified water provides accurate control of the plant water pH, minimizing the potential for corrosion of the plant equipment. An automatic central lubrication system and standby pumps in all the major circuits reduce maintenance problems and plant downtime.

After the cleaned coal is dried, all or part of the 1½ to 1¼ inch cleaned coal goes to a transfer buidling for automatic sampling. Another instrument in the plant, the Sortex Ash Monitor, delivers a continuous readout of the percentage of ash in the cleaned coal to the control room. The plant operator can then quickly identify a deviation from product specification and make the necessary changes in plant operation to bring it back into compliance.

FUTURE MARKETS

Pittston is optimistic about the future market for its cleaned coal. The company expects to sell cleaned coal to utilities using compliance coal for older boilers and for boilers converting from oil to coal, to foreign coal markets, to utilities that will burn coal-oil mixtures in their plants, and to industrial boilers.

Pittston has increased its sales of compliance coal from 4 million tons in 1979 to 5 million tons in 1980 and is counting on a rapid increase in the sale of compliance steam coal. According to a September 1980 *Business Week*, President Camicia foresees Pittston gaining a significant share of the compliance-coal market, which is expected to expand from a fledging market in the United States now to 25 or 30 million tons per year within the next two to three years.

K.H. Reardon, an analyst with A.G. Becker, Inc., in Chicago, states that Pittston will be able to take a larger share of the market for coal that will be used by utilities to meet sulfur dioxide standards without scrubbing. This market may include customers with older boilers and boilers switching to coal from oil or natural gas, because Pittston has preparation plants near all of its mines ''that can separate grades of coal carefully to provide uniformity.''

The future of the compliance-coal market is uncertain, however. Analyst Joel Price of Dean Witter Reynolds, Inc., suggests that when power plants now under construction "come on-stream in 1984 to 1985, that will be the end of the compliance coal market," as all newly constructed plants are now required by federal law to use flue gas desulfurization (FGD). The utilities that install scrubbers would not necessarily need this higher-priced, cleaner coal, although it may provide savings in FGD operating and maintenance costs.

Pittston is also pushing for increased exports of cleaned, high-quality steam coal. It is now bidding on a long-term contract with a European utility for 1 million tons per year; this would be the company's first steam-coal contract for export. Because the demand for all metallurgical coal for steel making is only expected to grow by 2 to 3 percent per year, the company does not expect to significantly increase its sales of such coal for coking to foreign buyers. However, Pittston now has the leading share of the coking-coal export market and expects to maintain it.

As noted earlier, Pittston sees a market for its cleaned, low-sulfur coal in coal-oil mixtures for utilities. The large and numerous paper companies in the Northeast are also viewed as potential buyers of the company's coal.

Waste Disposal

Pittston disposes of the 5,200 tons of wastes per day from its Rum Creek operation in a valley landfill. The company expects this area to be large enough to hold the wastes produced for the next 40 years. No slurry wastes are produced by the plant, and thus there is no need for impoundments to contain them. According to Pittston, the disposal costs of wastes from Rum Creek are confidential.

Costs

Pittston would not provide any cost figures for its Rum Creek coal cleaning operations. (For cost figures given by other companies interviewed by INFORM, see table in the chapter Findings.)

FACTORS INHIBITING AND ENCOURAGING THE DEMAND FOR CLEANED COAL

The Pittston Company feels that railroad shortages and severe congestion at U.S. coal ports are the main factors inhibiting demand for coal, and therefore for cleaned coal. Loading and unloading facilities at ports are also overtaxed.

Lack of adequate export facilities is critical for Pittston, as over 50 percent of its coal is sold to foreign customers and it hopes to increase its steam-coal exports. In an attempt to mitigate its coal transport problems, Pittston is participating in a consortium with Consolidation Coal Company (a subsidiary of CONOCO) and A.T. Massey Coal Company to construct new port facilities. They have developed a plan to spend about $100 million to build a new export coal terminal at Hampton Roads, Virginia, to handle about 20 million tons of coal annually. (For information on A.T. Massey, see A.T. Massey profile.)

Environmental constraints on the use of coal are another factor noted by Pittston that inhibits the growth of preparation plants. Major constraints on coal burning are responsible for this fuel being the source of only 20 percent of the nation's energy, according to Pittston. Pittston is hoping the new administration will take action by amending the Clean Air Act and by enacting legislation to encourage utilities to convert from oil to coal.

However, Pittston believes that there are several factors that encourage the growth of preparation plants. From the coal buyer's point of view, cleaned coal offers more Btu's (more energy) per ton than the same coal raw. Coal cleaning also produces a final combustion product that has less ash and sulfur, which at times can enable utilities to meet regulatory ceilings set for sulfur dioxide. Since in most cases the utilities pay the cost of transportation, the savings in transportation costs resulting from the reduced tonnage of cleaned coal usually accrue to the utilities, not to the coal cleaning companies.

RUM CREEK PREPARATION PLANT

Beginning of Commercial Operation: March 1980

Construction Time: Three & one-half years

Architectural, Engineering and
Construction Firm: McNally Pittsburg Manufacturing Corporation

Total Tons Processed: 826 tons of raw coal per hour; 1.3 to 1.5 million tons per year

Cleaned Coal Tonnage:

    Design Capacity: 475 tons per hour; 1.6 million tons per year

    Operating Capacity: 500 tons per hour; tons per year NA

Operates 16 hours per day, five days per week, 50 weeks per year

Customers: Mountaineer Power Plant of the Appalachian Power Company (a subsidiary of American Electric Power Company)

Manpower Required for Plant Operation: One superintendent, seven foremen, one clerk, and 43 additional personnel

Efficiency of Plant Operations:

    Preventive maintenance requirements: 60 man-hours per week

    Average downtime: NA

Resources

    Energy rating: 20.6 Mw (27,600 horsepower)

    Water use: 650 gallons of recycled water are required to process a ton of coal; make-up water needs for this plant are provided as well

    Magnetite requirements: 250 tons per month

    Reagent requirements for froth-flotation units: American Cyanamid frother and fuel oil are used at the rate of 0.4 pounds per ton of raw coal processed. No recovery system is used

Wastes Generated: 325 tons per hour, or 5,200 tons per day (40% of raw coal feed to plant)

R & F COAL COMPANY
Cadiz, OH 43907

Coal Cleaning Plant Profiled:	R&F Preparation Plant Warnock, Ohio
Mines Served:	Eight strip mines in Ohio

Total Number of Cleaning Plants:	One
Coal Production and Percentage Cleaned:	3.2 million tons in 1980, 100% cleaned
Total Number of Mines:	Eight strip mines in Ohio
Estimated Reserves (1979):	74 million tons; Approximate sulfur content -- 3.0%
Customers (1980):	
Utilities:	100%
Financial Information - R&F Coal Company:	NA
Parent Company:	Shell Oil Company
Total Coal Sales (1980):	3.2 million tons
Parent Company Total Revenues (1980):	19,959 million
Net Income (1980):	$1,542 million

The R&F coal cleaning plant is notable because it uses automatic sampling equipment designed to analyze accurately the characteristics of both the feed coal entering the plant and the cleaned coal leaving the plant. This analysis is very useful for efficient cleaning of the feed coal when its characteristics are extremely variable—as they are at R&F. This equipment also permits vigorous quality control of the cleaned coal to be sold.

Eighty percent of the cleaned coal from the R&F preparation plant is delivered to Tennessee Valley Authority's Colbert generating station in Mussel Shoals, Alabama. The remaining cleaned coal is delivered to three or four other utility users.

The case of R&F Coal Company is just one of an increasing number in which oil companies are becoming involved in the production, processing and sale of coal. In July 1977 the holding company for R&F Coal

Company, the Seaway Coal Company, was purchased by Shell Oil Company for $65 million. R&F Coal Company recently made its first venture into coal cleaning, completing its preparation plant in August of 1978. A modernization of Shell's Ohio River terminal facilities, through which R&F's coal is transported to customers, was finished the same year.

R&F PREPARATION PLANT

The company wanted to install thermal dryers in its preparation plant so that it could process the ultra-fine-size (-200 mesh) coal particles. However, the plant is located in a county that has not met the Environmental Protection Agency's particulate standards; hence no new sources of particulate emissions can be built in this area. Thermal dryers emit substantial amounts of particulates. Without a dryer, however, the ultra-fine-size coal particles would be too wet for a utility to gain energy from the Btu's contained within.

Sampling equipment at R&F does a "proximate" analysis of the cleaned coal every 30 minutes. The average sulfur concentration and heat value in each of the four 5,000-ton cleaned-coal storage silos is calculated. Clean coal can then be blended from each silo so that the average sulfur concentration and the coal's ability when burned to meet a given emission standard (lb $SO_2/10^6$ Btu) will comply with contract specifications.

Effectiveness of the Cleaning Plant

Because the coals from these counties vary, R&F Coal Company could give only a broad range for the feed and cleaned coal's sulfur, ash, moisture and Btu content.

R&F Preparation Plant cleans coal from three different seams (Pittsburgh No. 8, Meigs Creek No. 9, Waynesburg No. 11) in four counties in Ohio (Belmont, Guernsey, Harrison and Jefferson). Sometimes these coals are blended before they are sent through the plant; at other times coal from only one seam is processed. Heavy-media vessels, heavy-media cyclones and hydrocyclones are incorporated in the R&F plant.

The most common cause of downtime at the plant is wet or cold weather, which causes difficulty in coal handling, as wet or frozen coal clogs the conveyors and sometimes the cleaning equipment. To correct this problem, R&F doubled the capacity of the crushers ahead of the plant. Now coal that formerly went into the plant in big slabs is crushed first.

Coal Characteristics	Raw Coal	Cleaned Coal
Sulfur (%) :	2.00 to 6.00	1.25 to 5.25
Ash (%):	12.00 to 13.00	8.00 to 9.00
Moisture (%):	6.00	8.00
Btu per pound:	10,500 to 12,000	12,000 to 14,000

Btu recovery in cleaned coal: NA
Weight reduction in cleaned coal: 16% to 20% (23% when Versar, Inc., calculated cleaned coal production from April 22, 1980, to June 6, 1980)*

*Versar, Inc., a consulting firm in Springfield, Virginia, tested the "Attenuation of Sulfur Variability by Coal Preparation" at the R&F plant under a contract for the Environmental Protection Agency.

Waste Disposal

R&F Preparation Plant, which generates approximately 4,000 tons of waste per day (based on a 20 percent reduction in the weight of the raw coal and a 20-hour operating day), had to obtain the approval of four different agencies for its waste-disposal system. At present, the coarse refuse from the plant is being used to build a dam to hold water needed to provide for any loss in the plant's water. Fine refuse, which exits from the plant as 30 percent waste and 70 percent water, is disposed of in a slurry pond. Fines compacted in place serve to make the pond bottom impervious to acid leaching. Water is decanted from the slurry pond to the plant make-up water supply, thus closing the plant's water circuit.

After the dam is completed, a site to the north of the plant will serve as a disposal area for fifteen years. To comply with Surface Mining Control and Reclamation Act regulations, the refuse must be used to make a 100-foot-high hill to blend in with the existing ridge, and must be revegetated. A second disposal area is available to R&F when the first is filled.

According to William E. Spiker, Vice President of Administration at R&F Coal Company, leachate from preparation plant wastes is more toxic than leachate from mining wastes because the impurities such as sulfur are more concentrated. The slurry pond was built to catch any acid that might leach from the coarse-coal refuse site. The potential for environmental problems from acid leaching is further minimized at the R&F plant because the disposal sites are built above a rock bed.

FUTURE MARKET

Spiker believes coal cleaning has an excellent future because of the economics of coal use, as well as the technology's ability to reduce a coal's sulfur content. R&F sees its future market for clean coal as utilities. "Our coal (its quality) cannot compete for the metallurgical market," states Spiker, "and synfuels is at least a decade away." R&F sees large coal companies as the future owners of preparation plants, because coal cleaning is a capital-intensive business. Larger coal companies and those affiliated with oil companies generally have more funds available for the large capital expenditures needed for a preparation plant. (The R&F plant cost $22 million in 1978, plus an additional $8 million for front- and back-end loading equipment and cleaned-coal storage facilities.) Coal cleaning plants will be designed and built to clean even the smallest size coal particles, Spiker asserts, because of coal's increasing value.

R & F PREPARATION PLANT

Beginning of Commercial Operation:	August 1978
Construction Time:	22 months
Architectural, Engineering and Construction Firm:	Roberts & Schaefer
Total Tons Processed:	1,000 tons of raw coal per hour (800 tons of cleaned coal per hour produced); operates 20 hours per day, five days per week, or until a weekly quota of 50,000 tons of clean coal is produced
Customers:	80% of the coal to TVA's Colvert generating station in Mussel Shoals, Alabama. Rest to three or four other utility customers

Manpower Required for
Plant Operation: NA

Efficiency of Plant Operations

    Preventive Maintenance
    Requirements: Four hours per day

      Average Downtime: NA. However, attributed to wet
                                         or cold weather. To minimize this
                                         downtime, R&F is now involved in
                                         a winterization project (refer to
                                         text for details)

Resources

    Energy Rating: NA

    Water Use: Operates on a closed water loop
                                             basis; has a dam containing 30
                                         million gallons of water for reuse

    Magnetite Requirements: NA

Wastes Generated: 16% to 20% of raw coal feed,
                                         according to R&F

                                         23% of raw coal weight when
                                         Versar, Inc. calculated production
                                         from April 22, 1980 to June 6, 1980

Costs:

Capital Costs of Plant (1978): $22 million, with an additional $8
                                         million for front and back-end load-
                                         ing equipment and cleaned-coal
                                         storage facilities

Annual Operating and
Maintenance Costs: NA

Costs of Resources:

    Energy: NA

    Water: NA

    Magnetite: NA

Waste Disposal Costs: NA

Price Charged Per Ton
of Cleaned Coal: NA

Transportation Cost Per
Ton of Coal: NA

Financing Source for Plant: Funds from parent company

Utilities Owning Preparation Plants

AMERICAN ELECTRIC POWER, INC.
180 East Broad Street
Columbus, OH 43215

Coal Cleaning Plant Profiled:	Helper Preparation Plant Helper, Utah
Mine Served:	Price Mine, No. 3 Helper, Utah
Power Plants Profiled:	Mountaineer Power Plant New Haven, West Virginia
	Tanner's Creek Generating Station (near) Greendale, Indiana

Coal Use (1980):	42,337,000
AEP System Coal Production:	Underground production (66.6%) 9,357,409 tons
	Surface mine production (33.4%) 4,700,592 tons
	Total production 14,058,001 tons
AEP System Coal Deliveries:	Purchased from suppliers (68.8%) 30,318,000 tons
	Delivered from AEP mines (31.2%) 13,772,000 tons
	Total deliveries 44,090,000 tons

Total Number of AEP-
Owned Preparation Plants: Nine

Total Electrical Capacity (1980): 22,798 Mw

    Sources: 19 coal stations
                                   (87.37% of power production)

                                   One nuclear station
                                   (11.84% of power production)

                                   16 hydrostations
                                   (0.72% of power production)

                                   Four gas turbine plants
                                   (0.07% of power production)

                                   Five diesel plants
                                   (less than 0.01% of power production)

    Peak Load (1980):   19,488 Mw

    Reserve Capacity: 14.5%

Growth of Electrical Demand: Up 6.3% from 1979 (in Kwh)

Service Area: 47,872 square-mile area; over 3,000
                                   communities in Indiana, Kentucky,
                                   Ohio, Michigan, Tennessee, Virginia
                                   and West Virginia. Estimated pop-
                                   ulation of 7,006,000.

Number of Customers (1980):   2,487,888

Regulatory Commission(s):   Indiana Public Service Commission
                                   Kentucky Public Service Commission
                                   Ohio Public Utilities Commission
                                   Michigan Public Service Commission
                                   Tennessee Public Service Commission
                                   Virginia State Corporation Commission
                                   West Virginia Public Service Commission

Revenues (1980):   Total Operating Revenue
                                   $3.76 billion

Rates (1980):   Residential: 4.36¢/Kwh
                                   Commercial: NA
                                   Industrial: NA
                                   Average: 3.53¢/Kwh

The AEP Corporate Structure:

    Appalachian Power Company: Cedar Coal Company, Central Appalach
                                   Coal Company, Southern Appalachian C
                                   Company, Central Operating Company,
                                   Kanawha Valley Power Company

    Columbus & Southern Ohio
    Electric Company: Simco, Inc.

    Indiana & Michigan
    Electric Company: Blackhawk Coal Company, Price River
                                   Coal Company

Kentucky Power Company

Kingsport Power Company

Michigan Power Company

Ohio Power Company: Central Ohio Coal Company, Southern
                                    Ohio Coal Company. Windsor Power
                                    House Coal Company. Cardinal
                                    Operating Company.

Wheeling Electric Company

American Electric Power Company (AEP),* the largest electric power producer and coal consumer in the United States, is the owner of nine preparation plants, including the Helper Preparation Plant in Helper, Utah. The Helper plant is one of the most sophisticated facilities in the western United States. Heavy-media and froth-flotation equipment is used with a minimum of water consumption, conserving a resource valuable in arid Utah. AEP ships cleaned coal from this plant to its subsidiary in Indiana, Indiana and Michigan Electric Company's Tanner's Creek generating station.

Also profiled here is AEP's new Mountaineer Power Plant in New Haven, West Virginia, which receives "compliance" coal from Pittston's Rum Creek Preparation Plant, thus enabling AEP to meet sulfur dioxide emission standards. (For information on The Pittston Company and its Rum Creek plant, see The Pittston Company profile.)

AEP is a holding company that owns eight utilities in the east-central United States. In 1980 the company produced 111.1 billion Kwh of electricity, the largest amount of electricity produced by any American utility. Although it operates a large nuclear facility on Lake Michigan (the Donald C. Cook Nuclear Plant) and several hydropower projects, AEP produces over 87 percent of its electricity at its 19 coal-fired power plants.

With its 16 underground and 11 surface mines in Ohio and West Virginia, and its two underground mines in Utah, AEP was the nation's tenth largest coal producer in 1980. More than 83 percent of AEP's 1.8 billion tons of total recoverable eastern coal has a sulfur content greater than 2 percent. By contrast, AEP's recently acquired western reserves

---

*AEP did not provide information for this study, citing "the enormous expenditure of staff time [needed] to properly collect all the pertinent data." Therefore, the AEP profile is not as thorough as profiles of other companies.

contain 1.5 billion tons of recoverable coal having a sulfur content of 1 percent or less.

AEP claims that it remains committed to nuclear power. However, the company expects to become increasingly "coal-oriented" in the 1980s. Almost 98 percent of planned new generating capacity will burn coal. As it expands its use of coal, AEP states that it will attach "great importance to the quality of coal that we use in our plants in order to maximize boiler efficiency and availability and to decrease air pollution."

Coal Cleaning Policy

Since the adoption of the 1977 Amendments to the Clean Air Act, AEP has attempted to meet sulfur dioxide emissions standards by treating coal before combustion. Most of AEP's recent sulfur dioxide abatement attempts have involved coal cleaning. The company has "steadfastly refused to install scrubbers" to further reduce sulfur dioxide emissions. However, it has recently completed a $660 million retrofit program for installing electrostatic precipitators to control particulates at its generating stations.

AEP own nine coal preparation plants located at its eight captive mines in Ohio, Utah, and West Virginia. In 1979 construction of AEP's first demonstration coal cleaning plant using the Otisca Process was completed (see Otisca profile). In 1980 Windsor Power House Coal Company (an AEP subsidiary) began building a new coal preparation plant at its mine near Wheeling, West Virginia. A major addition to the Southern Ohio Coal Company's Meigs Mine Preparation Plant (another AEP subsidiary) was also begun in 1980. This facility will be fully operated by computers.

Sources of Coal and Coal Use

AEP owns, controls or has surface rights to more than 3.3 billion tons of recoverable coal reserves in Appalachia and the West, and has numerous subsidiaries that mine, transport and prepare this coal for its power plants. Because of its vertically integrated corporate structure, AEP was able to obtain approximately 30 percent of its coal from its own subsidiaries between 1975 and 1980; between 1975 and 1979, AEP obtained 55 to 61 percent of its coal through long-term contracts and 8 to

19 percent through spot-market purchases.

AEP's control over its coal supply is not confined to ownership of 29 mines and coal reserves. In addition, AEP subsidiaries own or lease many types of transportation equipment including 3,131 railroad hopper cars, 484 river barges, and 25 towboats. Recently the company completed construction of a rail-to-barge transfer terminal on the Kanawha River.

AEP indicates that it has been able to minimize transportation costs by locating most of its coal-powered generators at mine sites.

HELPER PREPARATION PLANT AND TANNER'S CREEK GENERATING STATION

INFORM chose to profile the Price River Coal Company's Helper Preparation Plant (an AEP subsidiary) in Helper, Utah, because it is an exceptionally sophisticated plant for the western United States. As noted earlier, the plant uses heavy-media vessels, water-only cyclones, heavy-media cyclones, and froth-flotation devices to clean raw coal. The Helper plant's water-control system cleans and recycles all process water, which is scarce in this desert region.

AEP's Helper Preparation Plant was completed in 1977 at a total cost of $10 million. With a feed rate of 1,250 raw tons per hour, the plant produces 1,100 tons of cleaned coal per hour, all of which is shipped to Indiana & Michigan Electric Company's Tanner's Creek Generating Station in Indiana. The quality of raw feed to the plant (its sulfur, ash, moisture and Btu content) cannot be compared to the quality of the cleaned coal, since AEP would not participate in INFORM's study. Information about the plant's efficiency, resource requirements, and waste disposal system is also not available.

AEP would not provide INFORM with specific information on the Tanner's Creek generating station such as the specifications set for the coal delivered to the power plant, the boiler's operating experience with cleaned coal, or any cost savings attributed to burning cleaned coal in the plant.

Mountaineer Power Plant

INFORM was unable to secure information on the impact of cleaned coal on power generation at the new 1.3 million Kw Mountaineer Plant owned by an AEP subsidiary, the Appalachian Power Company. Located on the Ohio River at New Haven, West Virginia, the plant burns cleaned coal from several sources including The Pittston Company's Rum Creek Preparation Plant in Logan County, West Virginia. (For more information on The Pittston Company and its Rum Creek Plant, see The Pittston Company Profile.)

DUQUESNE LIGHT COMPANY
435 Sixth Avenue
Pittsburgh, PA 15219

Coal Cleaning Plant Profiled:	Warwick Preparation Plant Greensboro, Pennsylvania
Mine Served:	Warwick Mine Greensboro, Pennsylvania

Coal Use (1979):   3.7 million tons for three company-owned powerplants

Sources:

    Captive Production: 24%
    Long-Term Contracts: 27%
    Spot Market: 49%

Coal Production at Captive Mine (Warwick): (tons per year)		
1979:	927,890
1978:	699,123
1977:	727,779
1976:	1,079,122

Estimated Recoverable
Reserves (12/31/79):   29.9 million tons at Warwick Mine

Total Number of Duquesne-
Owned Preparation Plants: One

System Generating Capability
(1980):   3,179 Mw

Sources: Three company-owned coal stations
Five jointly-owned coal stations
(92% of power production)

Two jointly-owned nuclear stations
(3% of power production)

	One company-owned oil station (1% of power production)
	Outside purchases (4% of power production)
Peak Load (1980):	2,474 Mw
Reserve Capacity:	22%
Growth of Electrical Demand:	(0.2%) from 1979-80 (decline)
Service Area:	800 square mile area that includes the City of Pittsburgh and two surrounding counties
Number of Customers (1980):	552,502
Regulatory Commission:	Pennsylvania Public Utility Commission
Total Operating Revenue:	$689,465,000
Fuel Purchases:	$212,672,000 (31% of total operating revenue)

Rates (1980):

Residential:	6.83¢/Kwh
Commercial:	5.22¢/Kwh
Industrial:	3.99¢/Kwh
Average:	5.35¢/Kwh

The Warwick Preparation Plant, owned by Duquesne Light Company, ships cleaned coal to the company's three coal-fired power plants near Pittsburgh, Pennsylvania. This cleaning plant is one of two in the United States that use 24-inch water-only cyclones to clean fine coal. The cyclones enable Duquesne Light to save both space and money at Warwick. The Warwick plant was also designed to recover a high percentage (96 percent) of the Btu content from the raw coal, saving additional money, since few Btu's are lost in the refuse.

Duquesne Light cleans coal because cleaning 1) reduces the sulfur content of the raw coal from the company-owned Warwick Mine, decreasing sulfur dioxide emissions at Duquesne's power plants, and 2) removes high percentages of ash and mining wastes, and thus raises the Btu value of the fuel to meet the power plant's fuel specifications.

Duquesne Light's Coal Cleaning Policy

A mixture of raw and cleaned coal is burned at all Duquesne Light's generating stations. Often, cleaned coal is used as part of the fuel mix at the plants in order to help meet sulfur dioxide standards, although this is not the only reason Duquesne Light uses it. Cleaned and uncleaned lower-sulfur coals can be blended with higher-sulfur coals at any generating station in order to reduce sulfur emissions if necessary. Duquesne Light claims that the company does not know what percentage of the coal burned at each station is cleaned.

The following table lists generating stations owned entirely by Duquesne Light.

Station	Unit	Size of Unit (Mw)	In-Service Date	Fuel Pre-cleaned(%)
Cheswick	1	570	1970	Variable
Elrama	1	100	1952	
	2	100	1953	
	3	112	1954	Variable
	4	175	1960	
Phillips	1	35	1942	
	2	38	1943	
	3	60	1943	
	4	60	1943	Variable
	5	60	1949	
	6	140	1954	

WARWICK PREPARATION PLANT

The Warwick Preparation Plant, owned and operated by Duquesne Light, cleans all the coal produced at the nearby Warwick Mine. The plant is situated on the Monongahela River, 83 miles south of Pittsburgh, at the site of an old coarse-coal cleaning facility that has been decommissioned.

The Warwick Preparation Plant is characterized by two special features. First, the plant is designed to retain in the cleaned coal a large percentage (96 percent) of the Btu's in the raw coal. Second, it uses 24-inch water-only cyclones to clean the raw coal that is less than 3/8 inch in size. As of November 1980, only one other plant in the United States used this equipment to clean fine coal.

The use of 24-inch water-only cyclones ahead of a heavy-media vessel has several advantages:

1. The cyclones require little space.

2. Significantly less raw-coal screening equipment is required with this design, lowering equipment costs.

3. Water-only cyclones can remove most of the cleaned coal from the 3/4 inch to 0 fraction (which may be two thirds to three quarters of the raw coal).

4. Fewer heavy-media vessels are required, since the cyclones have reduced the amount of coal needed to be cleaned by heavy media. Additional money is saved, since fewer pieces of heavy-media equipment are needed and less magnetite is used.

5. The quarter-inch wastes from the 24-inch water-only cyclones are recleaned in 12-inch water-only cyclones, which are set so that very few Btu's are in the reject. This results in a high percentage of Btu recovery for the whole plant.

Effectiveness of the Cleaning Plant

Due to the high ash and low Btu content of the raw coal, Duquesne Light must clean the coal mined at Warwick, so that the fuel will meet generating station specifications. About 35 percent of the raw coal is removed as reject, while 96 percent of the Btu value of the raw coal is retained in the cleaned coal.

The following table compares the characteristics of the raw and the cleaned coal:

Effectiveness of the Cleaning Plant

Coal Characteristics	Raw Coal	Cleaned Coal	Decrease/Increase
Sulfur (%):	2.3	1.9	-17.40
Ash (%):	32.0	15.9	-50.30
Moisture (%):	3 to 4	6.3	+57.50 to +110
Btu per pound:	8,000	11,500	43.75%

Btu recovery in cleaned coal: 96%

Source of Raw Coal

The underground Warwick Mine is Duquesne Light's sole captive mining operation. The utility owns and operates the mine, which produces all its coal for the Warwick Preparation Plant. High-volatile bituminous coal from the Sewickley seam is now mined at Warwick. Raw coal is transported from the mine on a five-mile overland conveyor belt to a storage facility near the preparation plant, from which it is moved again via conveyor belt to the plant.

Wastes

Preparation plant wastes are layered at a disposal site located near the plant. Duquesne Light estimates that this site is large enough to contain wastes generated for the next 30 years. The utility sees the more efficient recovery of fine coal from plant wastes as the ultimate solution for most of the problems encountered in preparation plant waste disposal. The company would not comment on its success or failure in meeting federal regulations governing preparation plant operations and waste disposal. However, a spokesman did state that there is no practical way to meet the federal Mining Enforcement and Safety Administration noise regulations at the Warwick plant.

Duquesne Light was one of the two plant owners in INFORM's sample that provided waste-disposal costs. At the Warwick plant, it cost $1.467 in 1979 to dispose of the wastes remaining after one ton of cleaned coal was produced (see Findings on waste disposal and waste chapter in Appendix).

Capital and Operating and Maintenance Costs

Duquesne Light reported that the capital cost of the Warwick preparation plant was $15 million (in 1979 dollars). Annual operating and maintenance costs were not provided by the company.

CHESWICK, ELRAMA AND PHILLIPS
GENERATING STATIONS

All the coal mined and cleaned at the Warwick site is burned at the Cheswick, Elrama and Phillips generating stations in western Pennsylvania (in Springdale, Elrama and Wireton, respectively), which are owned and operated by Duquesne Light.

Emission Standards

Cleaning coal from the Warwick Mine, to decrease its sulfur content and variability, is especially important because the U.S. Environmental Protection Agency (EPA) has designated the areas in which Duquesne Light's generating stations are located "non-attainment areas" for sulfur oxides and particulates under the Clean Air Act. This designation has resulted in stringent sulfur dioxide standards, requiring the company's Elrama and Phillips power plants to meet an SO_2 emission standard of 0.6 pounds of SO_2 per million Btu. No averaging time is granted for meeting this standard. Thus the use of coal with a higher sulfur content may not be compensated for later by the use of coal with a lower sulfur content. New state and local implementation plans are also being formulated in response to the designation. Consequently, additional pollution-control equipment and/or changes in plant operations may be required at Duquesne's generating stations.

The Elrama and Phillips stations employ flue gas desulfurization (FGD) systems in order to meet sulfur dioxide standards.* The company had attempted to meet the former Allegheny County sulfur dioxide standards of 0.6 pounds SO_2 per million Btu at the Cheswick station by burning partially cleaned low-sulfur coal without the use of FGD equipment.† However, EPA issued a "notice of violation" to the company under Section 120 of the Clean Air Act for non-compliance at the

---

*See INFORM's flue gas desulfurization study for more information about Duquesne Light Company's use of scrubbers.

†In February 1980 Duquesne Light signed a six-year contract with Penn Allegheny Coal Company for 600,000 to 720,000 tons annually of deep-mined, 1.1 percent sulfur coal, explicitly for the Cheswick station. Part of the coal mined and cleaned at Warwick is also burned at Cheswick.

Cheswick station. Duquesne Light proposed a modification in the Allegheny County sulfur dioxide standards that would allow the continued burning of partially cleaned low-sulfur coal at Cheswick in lieu of installing scrubbers. In October 1981, the modification proposal of the Pennsylvania State Implementation Plan (SIP) under the Clean Air Act was given final approval by EPA. Cheswick's sulfur dioxide standard has been raised from 0.6 to 2.8 pounds per SO_2 per million Btu.

Coal Specifications

The following table lists the specifications that must be met by captive and outside coal suppliers, with either cleaned or high-quality raw coal:

	Cheswick	Elrama	Phillips
Sulfur (%):	1.4 to 1.6	2.5	2.3
Ash (%):	11 (+)	18.0	15.5
Moisture (%):	6	5.7 to 5.8	5.5
Btu per pound:	12,300	11,450	11,800

Benefits and Problems of Burning Cleaned Coal

Duquesne Light sees several advantages in the use of cleaned coal at its generating stations. Coal cleaning can reduce the sulfur content of raw coal and thus the sulfur dioxide emissions when the fuel is burned. This enables the utility to use coal cleaning as part of its strategy for meeting air-quality standards. The company also believes that coal cleaning might reduce the load on scrubbers in stations where they are used. A lower-sulfur coal means that less sulfur needs to be removed from the flue gases, reducing operating costs and the volume of scrubber sludge generated. Other advantages of burning cleaned coal noted by Duquesne include lower transportation, ash-handling, and power station maintenance costs. Engineering and economic analyses of raw and cleaned coal combustion have been performed by the company, but were not made available to INFORM.

Duquesne Light did not note any problems associated with the burning of cleaned coal at its power plants.

Transportation

Coal is transported to Duquesne Light's generating stations by either barge or truck. Barge deliveries are made daily from Warwick; truck deliveries from outside coal suppliers arrive five days a week. (The Warwick Preparation Plant is less than 100 miles from each of the generating stations, and all of the coal shipments from the Warwick site are made by barge.)

The cost of transporting cleaned coal from Duquesne Light's preparation plant to its power stations was $0.80 per ton of cleaned coal. This is the lowest transportation cost figure in INFORM's sample of utilities. This low cost can be attributed to the use of barges to move the coal; rail and truck are usually more expensive modes of transportation.

Coal Costs

The average price of coal purchased by Duquesne Light in 1979, both from its own mines and from outside suppliers, was $30.84 per ton, or $1.316 per million Btu. However, cleaned coal prices were generally higher, depending on the sulfur content of the fuel. In 1980 the approximate prices per ton were $38.00 for deep-mined, low-sulfur cleaned coal, and $30.00 for surface-mined, average-sulfur cleaned coal.

Duquesne Light does not compute the price of raw coal from the Warwick mine. The company computes only the price of the cleaned coal that is shipped to its generating stations. A spokesman for Duquesne Light said that these cleaned coal prices "vary" and declined to quote specific prices. However, the spokesman did stress that captive mining operations are not independent of the "outside" coal market forces, and that Duquesne uses its captive mine and preparation plant "as a lever to get cheaper coal from the open market." The utility maintains that, being a coal producer as well as a coal buyer, it can compare its production costs with prices quoted by suppliers and use this information to its advantage during pricing negotiations.

FUTURE PROSPECTS

Duquesne Light views the Warwick Mine and Preparation Plant as integrated components of its overall coal-supply system. Therefore the company claims that problems in mine operation cannot be separated from those encountered in the preparation plant. Absenteeism and strikes have limited captive-coal production and have reportedly caused the Warwick Mine and Preparation Plant to operate at less than full capacity.

L.W. Johnson, Duquesne Light's Superintendent of Technical Services, summed up his thoughts concerning the role of government in the future use of coal cleaning by utilities in the following statement: "The role of government will increase...this is very sad." Another company spokesman concluded his thoughts on coal cleaning and other pollution-control technologies by suggesting that "one needs to be sure that environmental technologies are needed before they are built...society must have a return on the money it spends."

WARWICK PREPARATION PLANT

Beginning of Commercial Operation:	1974
Construction Time:	Approximately two years
Architectural, Engineering and Construction Firm:	Dravo Corporation
Total Tons Processed:	900 tons of cleaned coal per hour. Can process entire yearly output of Warwick Mine (927,890 tons in 1979). Manned to operate 14.5 hours per day, five days per week, 50 weeks per year, minus holidays
Users:	All cleaned coal used by Duquesne Light Company-owned power plants: Cheswick, Elrama and Phillips
Manpower Required for Plant Operation:	27-man crew, including four foremen, operates and maintains preparation plant and other Warwick surface facilities

Efficiency of Plant Operations

Preventive maintenance
requirements: 7.5-hour maintenance shift per
                              day; also, plant maintained when
                              not running

Average downtime: About 2% of the time that Duquesne
                              Light wanted plant to operate

Resources

Energy rating: NA

Water use: Closed water loop--125 gallons of
                              make-up water added per minute

Magnetite requirements: Approximately 1.2 pounds per
                              ton of raw coal for heavy-media
                              washers (requirement changes
                              with raw coal supply)

Wastes Generated: 1,885 tons when 3,500 tons
                              of clean coal are produced

COSTS

Capital Cost of Plant (1979): $15 million

**Annual Operating and
Maintenance Costs:**   NA

Cost of Resources:

Energy: $3,895.95 (from August 21,
                                       1980, through September
                                       22, 1980)

Water: NA

Magnetite: NA

Waste-Disposal Costs (1979): $1.467 per ton of cleaned coal

**Price Charged Per Ton
of Cleaned Coal:**   NA

**Transportation Cost per
Ton of Coal:** $0.80 per ton of cleaned coal
                                       (barge)

Financing Source for Plant: Funding for Warwick Preparation
                                       Plant came from sale of Duquesne
                                       Light Company stock and bonds

PENNSYLVANIA ELECTRIC COMPANY
1001 Broad Street
Johnstown, PA 15909

Coal Cleaning Plant Profiled:	Homer City Preparation Plant Homer City, Pennsylvania
Mines Served:	Helen and Helvetia Mines Homer City, Pennsylvania
Powerplant Profiled:	Homer City Generating Station Units 1-3 Homer City, Pennsylvania

Coal Use (1979):	14.9 million tons
Sources (1979):	"Dedicated" production, 29% Long-term contracts, 27% Spot market, 44%
Dedicated Mines: (for Homer City)	Helen Mining Company (subsidiary of North American Coal Corporation)
	Helvetia Mining Company (subsidiary of Rochester & Pittsburgh Coal Company)
Coal Production at Dedicated Mines (1979):	Helen, 990,329 tons Helvetia, 2,268,506 tons
Estimated Recoverable Reserves:	Helen, 33,000,000 tons Helvetia, 68,000,000 tons
Total Number of PENELEC- Owned Preparation Plants:	One
Total Electrical Capacity of Pennsylvania Electric Company (1980):	2,729 Mw
Sources of Electricity:	Five company-owned coal stations; one jointly-owned coal station (99% of power production)
	Three hydro/one pumped storage sta- tions (less than 1% of power production
	Three combustion turbine stations (gas and fuel oil) Three diesel peak load stations (less than 1% of power production)
	One jointly-owned nuclear station (Three-Mile Island - not operating)
Peak Load (Winter, 1980):	2,091 Mw

Reserve Capacity:	24%
Growth of Demand for Electricity:	5.2% from 1970 through 1980 (projected 1% annually through 1990)
Service Area:	17,600 square mile area that includes northwestern, northern, and south-central Pennsylvania
Number of Customers (1980):	512,000
Regulatory Commission:	Pennsylvania Public Utility Commission
Total Operating Revenues (1980):	$522,129,000
Fuel Purchases (1980):	$169,750,000 (33% of total operating revenue)
Rates (1980):	Residential, NA Commercial, NA Industrial, NA Average (1980), 4.57¢/Kwh
Parent Company:	General Public Utilities Corporation (GPU)
Affiliated Companies:	Jersey Central Power & Light Co. (JCP&L) and Metropolitan Edison Co. (Met-Ed)

The Homer City Preparation Plant is unique because it will supply two grades of cleaned coal to the Homer City Generating Station, located at the plant site. The use of two grades of coal will enable the Homer City station to meet environmental regulations without using flue gas desulfurization (FGD) equipment. One grade will provide Homer City generating units 1 and 2 with medium-sulfur coal that allows the units to meet Pennsylvania's sulfur dioxide emission standards for "older" boilers. The second grade, a low-sulfur coal, will be produced by the same preparation plant for the recently built Unit 3, which must meet the federal New Source Performance Standards (NSPS).

The plant is operated by Pennsylvania Electric Company (PENELEC); it is jointly owned (50 percent each) by PENELEC and New York State Electric and Gas Corporation (NYSE&G). PENELEC was incorporated in 1919 and is a subsidiary of General Public Utilities Corporation (GPU), the nation's fourteenth largest investor-owned electric utility.* The Homer City Preparation Plant is PENELEC's first effort in this field.

PENELEC's Coal Cleaning Policy

As with most of the utilities profiled in this study, PENELEC does not require the use of cleaned coal at its power plants. It sets specifications for sulfur and Btu content, and coal suppliers must provide fuel that meets these specifications, either by cleaning coal or by providing suitable raw coal. As a result of this policy, the percentage of cleaned coal burned at PENELEC's different stations varies, as shown in the following table:

Station	Unit	Size of Unit (Mw)	In-Service Date	Fuel Pre-cleaned(%)
Homer City	1	609	1969	100*
(50% owned)	2	609	1969	100*
	3	636	1977	100*
Shawville	1-4	628 (total)	1954-60	30
Seward	3-5	218 (total)	1941-57	100
Front Street	1-5	118 (total)	1917-53	0
Williamsburg	1	25	1944	0
Warren	1-2	85 (total)	1948-49	25

*When fully operating

Research and Development

The U.S. Environmental Protection Agency (EPA), the Electric Power Research Institute (EPRI), and the Homer City owners are sponsoring a three-year project to assess physical coal cleaning technologies

---

*The Three Mile Island nuclear accident has had a significant effect on GPU and PENELEC finances. The Pennsylvania Public Utility Commission ruled that the cost of cleaning up GPU's nuclear station could not be recovered from utility customers, which has threatened GPU with bankruptcy. As a result, all PENELEC construction activities have been shelved for the time being.

(mainly at Homer City)—their ability to satisfy sulfur dioxide standards and their applicability on a nationwide scale. The Homer City Generating Station will be studied to determine the impact of burning cleaned coal on the operation and economics of power generation.

PENELEC, NYSE&G, EPRI, Kaiser Engineers and the Empire State Electric Research Corporation are involved in the EPRI Homer City Preparation Plant Test Facility, which will measure the potential for cleaning different types of raw coal.

HOMER CITY PREPARATION PLANT

The Homer City Preparation Plant represents an advanced application of the physical coal cleaning technologies, in that it is designed to produce two grades of cleaned coal in each of two independent circuits. This will allow for a partial shutdown in the event of operating difficulties. In addition, the plant contains several innovative uses of conventional circuitry and equipment. The heavy-media cyclones, used to clean medium- and fine-size coal particles, clean coal sizes as small as 100 mesh, rather than the usual minimum of 28 mesh. The specific gravity of the heavy-media equipment has been lowered to 1.3 from the 1.4 or 1.5 that is normally employed, in order to insure extremely good product quality in the intensively cleaned coal. A new monitoring system has been designed to control specific gravity tightly in the "low-gravity" (specific gravity of 1.3) circuits, which aids in achieving maximum separation. Double-drum magnetic separators have been used to supplement the drain-and-rinse screens in order to better recover magnetite from the heavy-media equipment in the fine-coal circuits. Finally, thermal dryers are used to reduce the moisture content of the cleaned coal.

Effectiveness of the Cleaning Plant

The following table lists the specifications set by PENELEC and NYSE&G for raw and cleaned coal to be met by raw-coal suppliers and the preparation plant:

	Raw Coal	Medium Product	Deep-Cleaned Product	Medium Product: Decrease/Increase*	Deep-Cleaned Product: Decrease/Increase*
Sulfur (%):	3.4 maximum	2.35	1.81	-31.00	-46.90
Ash (%):	20.0 maximum	17.75	8.00	-11.25	-60.00
Moisture (%):	6.0 maximum	4.00	4.00	-33.00	-33.00
Btu per pound:	11,800 minimum	12,549	14,000	+6.35%	+18.64%

Btu recovery in cleaned coal: 95.4%
(Approximately 62% in medium product and 32% in deep-cleaned product

*PENELEC expects to achieve 34% and 75% sulfur removal in the medium-cleaned and deep-cleaned products, respectively, once the equipment retrofit activities are completed in February 1982. The company also expects to reduce the ash content of the deep-cleaned coal by 86%, raising its Btu content to 14,500 by 1982.

Preparation Plant History

The Homer City Preparation Plant was originally designed as a conventional physical coal cleaning facility, to provide a medium-sulfur and medium-ash cleaned coal product for the Homer City Generating Station Units 1 and 2. The planned addition of a new coal-fired unit to the Homer City station encouraged PENELEC and NYSE&G to convert to a "Multi-Stream Coal Cleaning System" (MCCS) at Homer City. The MCCS is designed to produce both a medium-sulfur (2.35 percent) and medium-ash (17.75 percent) coal for the older units (1 and 2) and an intensively cleaned (or "deep-cleaned") low-sulfur (1.81 percent) and low-ash (8.00 percent) coal for the newest Homer City generating unit (3). The MCCS circuitry had to be worked into the original preparation plant configuration, since the plant was already under construction when the design change was made in 1975.

The plant has been operated by Iselin Preparation Company, a subsidiary of Rochester & Pittsburgh Coal Company, since its initial start-up in 1977. The original washing circuitry was constructed to go into operation in stages throughout the fall of 1978. However, after more than a year of equipment tests and experiments with operating conditions, a decision was made to redesign the plant equipment and configuration.

On June 27, 1980, the plant was shut down to begin major equipment and plant modifications, owing to difficulties in achieving the original cleaning goals. Several consultants attribute the plant's difficulties to the decision to incorporate the MCCS into the original plant configuration, instead of halting construction and redesigning the plant around the MCCS.

Sources of Raw Coal

Eighty percent of the coal cleaned at Homer City is from the underground Helen and Helvetia dedicated mines, and 20 percent is purchased from outside coal companies on the spot market. PENELEC provides financial support to private coal companies located near the Homer City site to develop mines that are "dedicated" to supplying coal to Homer City for the life of the preparation plant and generating station. These mines supply medium-volatile Indiana County bituminous coal from the Upper Freeport "E" seam and the Lower Freeport "D" seam.

Problems with the Preparation Plant

Translating the conceptual design of the Homer City Preparation Plant into an operating facility has proved to be a difficult task for PENELEC and NYSE&G. The plant has encountered extensive problems since operations began in 1977—design errors, the misapplication and poor layout of equipment, inadequately trained operating personnel, and insufficient numbers of qualified start-up and design personnel. Problems have occurred with critical equipment such as the heavy-media cyclones, magnetic separators, and water-clarification system. As previously noted, portions of the plant are being redesigned and rebuilt by several consulting and engineering/construction firms, and an operating and maintenance training program has been instituted to address the personnel problems.

Wastes

Most of the wastes generated at the Homer City Preparation Plant are trucked from the plant to a landfill about one mile from the plant site. A collection pond is also used, to retain fine refuse for later disposal in the same landfill. PENELEC reports that collection ponds are "treated and maintained as required" by federal and state regulations. PENELEC did not provide information on the cost of waste disposal at the Homer City plant.

The Homer City Preparation Plant has encountered regulatory difficulties over fugitive dust emissions. Fine coal particles have presented a dust problem in the conveyor system and at the generating station. Dust-control sprays and covers over the conveyor belts are being installed to correct this condition. Thermal dryers have been outfitted with scrubbers to control particulate emissions from the coal-fired dryer furnaces.

Capital and Operating and Maintenance Costs

The capital cost of the Homer City Preparation Plant is estimated by PENELEC to be $97,200,000 (to the end of 1981). (See cost table at the end of this profile for a breakdown of capital costs.) This is the most expensive plant in INFORM's sample of cleaning facilities. A significant percentage of the capital costs (39 percent) have been for plant modification, as discussed above. PENELEC estimates Homer City's annual operating and maintenance costs at $13 million per year. This figure is also the highest in INFORM's sample.

HOMER CITY GENERATING STATION

The Homer City Generating Station is the major user of cleaned coal in the PENELEC system. This facility burns over one percent of the coal used to generate electricity in the United States. Most of the cleaned coal for Homer City comes from the Homer City Preparation Plant.

The Homer City Generating Station comprises three units that are designed to burn medium-sulfur coal (Units 1 and 2) and low-sulfur coal (Unit 3). Units 1 and 2 burned raw coal until sulfur dioxide regulations, promulgated in Pennsylvania in March 1972, limited sulfur dioxide emissions from these sources. Soon after, a decision was made to construct an on-site cleaning plant to treat raw coal for the existing units. In August 1975, after studying several environmental control options for a new Homer City generating unit (including an FGD system), PENELEC and NYSE&G chose to construct the Homer City Preparation Plant. The specifications set by PENELEC for coal burned at the Homer City Generating Station have been presented in the preceding section on the preparation plant. Units 1 and 2 meet the Pennsylvania sulfur dioxide emission standard for "older" boilers (3.7 lb $SO_2/10^6$ Btu on a 30-day running average); and according to PENELEC, Unit 3 will meet the federal NSPS (1.2 lb $SO_2/10^6$ Btu on a three-hour averaging period), once the preparation plant is successfully retrofitted.

Benefits and Problems of Burning Cleaned Coal

James Tice, PENELEC's Manager of EPA/EPRI Research and Development, sees cleaned coal as a means of holding down consumer electricity rates. Burning cleaned coal increases the availability of generating units, since the lower ash content decreases boiler outages related to slagging and fouling, and thus increases the power output of the boilers. This "extra" generating capacity can help forestall the construction of additional generating units, since existing facilities can produce more electricity more of the time.

Boiler performance has improved in Homer City Station Units 1 and 2 since PENELEC switched to burning cleaned coal from run-of-mine coal, and the boilers have been able to increase their generating capacity. In the past PENELEC encountered problems with waste buildup in the boilers and boiler efficiency, due to slagging and fouling in Units 1 and 2. At first, boiler availability was improved by slowing the rate at which coal was fed into the boilers, resulting in less fouling, slagging and erosion. However, the electrical output of the station fell by 70 Mw to 80 Mw, since less fuel was entering the boilers. The electrical output of the station increased, however, when cleaned coal was burned instead of run-of-mine coal; the average boiler rating rose by 25 Mw, partially offsetting the effects of decreasing the fuel supply.

Certain mineral materials, which become more concentrated in intensively cleaned coal because of selective ash removal, can cause slagging and fouling difficulties in some boilers. However, boiler studies conducted by Babcock & Wilcox, the Australian Coal Research Group, and Pennsylvania State University, prior to the use of cleaned coal at Homer City, indicated that combustion of intensively cleaned, low-sulfur and low-ash coal would not lead to slagging and fouling difficulties in Unit 3. The experience so far at Unit 3 indicates that the study results were correct.

James Tice also described the positive effects of cleaned-coal use upon several other power plant activities. Cleaned. coal is easier to pulverize than raw coal, states Tice, which suggests lower operating costs for pulverizers being fed cleaned coal as opposed to raw coal. The cost of boiler-ash disposal is also reduced with cleaned coal, due to decreases in the amount of boiler ash handled.

However, the use of cleaned coal raises the cost of coal handling and storage at Homer City, because the fine coal sticks to the chutes and belts. Also, dust from fine coal particles has become a problem, requiring the use of dust-collection and dust-suppression equipment (sprays, and covers over conveyor belts).

Tice says that when medium-sulfur cleaned coal is burned, electrostatic precipitator (ESP) operations cost less than when raw coal is burned, due to a drop in fly ash production. However, PENELEC reports little experience in burning intensively cleaned, low-sulfur and low-ash coal in Unit 3—coal which may adversely affect ESP operations. The "resistivity" of the fly ash is altered by the use of cleaned coal; this may reduce the electrical charge on the fly ash particles and reduce the ability of the ESP to attract and collect them.

PENELEC has not estimated the cost of using cleaned coal, beyond general qualitative descriptions of the benefits and problems of its use, at Homer City.

Transportation

The coal cleaned at the preparation plant is moved to the generating station by a half-mile-long conveyor belt. Coal purchased from outside suppliers is shipped to the generating station by truck. PENELEC and NYSE&G may be committing additional reserves to the Homer City Generating Station as future needs indicate. Transportation of coal by rail and/or slurry pipeline will be considered as alternatives at that time. PENELEC did not provide data on the cost of shipping coal to the Homer City plant.

Coal Costs

At present, the cost of coal purchased by PENELEC on the open market for the Homer City Generating Station averages $27.74 per ton. PENELEC does not sell any of the cleaned coal produced at its preparation plant. However, the cost of the medium-cleaned and intensively-cleaned coal that will be produced in 1982 at the Homer City Preparation Plant for use at the power plant has been estimated by the company at approximately $2.00 and $6.00 per ton, respectively, above the price of raw coal.

FUTURE PROSPECTS

According to James Tice, innovative coal cleaning technologies can be encouraged by environmental regulations. The Homer City MCCS is an example of the development of a new plant design in response to strict sulfur dioxide regulations. Although FGD was considered by PENELEC as a sulfur dioxide control technology, the advanced MCCS plant was chosen because original cost estimates indicated that it would be cheaper to build and operate. However, most of PENELEC's generating stations are older facilities that are subject to Pennsylvania sulfur dioxide regulations, which are less stringent than the federal NSPS. FGD and advanced coal cleaning are not necessary for such stations, but Tice sees a role for less rigorous coal cleaning in order to bring locally mined coals into compliance for use in the older plants.

HOMER CITY PREPARATION PLANT

Beginning of Commercial
Operation

    Medium-cleaned coal circuit: December 1977

    Deep-cleaned coal circuit
    (intensively cleaned):   Fall 1978 (only sporadic
                                 operations since then)

Construction Time

    Medium-cleaned coal circuit: Approximately two years

    Deep-cleaned coal circuit: Approximately three years

Architectural, Engineering and
Construction Firm

    Original Plant: Heyl & Patterson, Inc.

    Retrofit: Ebasco Services, British Mining
Consultants, Kaiser Engineers

Total Tons Processed

    Design Capacity: Two independent circuits--
a total of 1,200 tons of raw
coal per hour

                                        960 tons of cleaned coal per
hour, 4,760,000 tons per year
(1.3 million tons of deep-cleaned
coal and 3.5 million tons of
medium-cleaned coal)

    Operating Capacity: 485 tons of cleaned coal per hour
(a total of 535,000 tons) from
January through May 1980--
only one circuit in operation

User: PENELEC/NYSE&G Homer City
Generating Station

Manpower Required for
Plant Operation: 106 operators to run plant,
plus eight engineers and six
chemists for start-up

Efficiency of Plant Operations

    Preventive maintenance
    requirements: 88 hours per week (44 per
circuit) from January through
May 1980

    Average downtime: High, due to problems in
start-up of plant

Resources

    Energy rating: 12 Mw (20 Mw at full capacity)

    Water use: Closed water loop--16,000
gallons per minute at full
capacity (figure for make-up
water not available)

    Magnetite requirements: 2 to 3 pounds per ton of
medium-cleaned coal,

                                        5 to 7 pounds per ton of
deep-cleaned coal.

Wastes Generated: 5,000 tons per day at full
capacity

COSTS

Capital Cost of Plant

Deep-cleaned coal section:	$32,300,000
Medium-cleaned coal section:	$19,800,000
Modifications:	$38,100,000
Truck dump:	$ 7,000,000
Total:	$97,200,000 (to end of 1981)

Annual Operating and
Maintenance Costs:

$13,000,000 per year
(current estimate)

Cost of Resources
(Energy, Water, Magnetite):

NA

Waste-Disposal Costs:

NA

Price Per Ton of Cleaned Coal
Above Price for Raw Coal:

$6 in 1982 for deep-cleaned
product (estimated)

$2 in 1982 for medium-cleaned
product (estimated)

Transportation Cost per
Ton of Coal:

NA

Financing Source For Plant:

50% each: PENELEC and NYSE&G.
PENELEC funding from bond sales
and reinvestment of company
profits

PENNSYLVANIA POWER and LIGHT COMPANY
Two North Ninth Street
Allentown, PA 18101

Mining Subsidiary:	Pennsylvania Mines Corporation (PMC) P.O. Box 367 Ebensburg, Pennsylvania
Coal Cleaning Plant Profiled:	Greenwich Collieries Preparation Plant Greenwich, Pennsylvania
Mines Served:	Greenwich Collieries No. 1 and No. 2 Mi Greenwich, Pennsylvania
Power Plant Profiled:	Montour Station, Units 1 and 2 Washingtonville, Pennsylvania

Coal Use (1979):	9,581,653 tons
Sources:	"Captive" production (PMC), 31% Long-term contracts, 25% Spot market, 44%
Subsidiary Mines: (operated by PMC)	Greenwich, 2.0% sulfur (average) Rushton, 2.2% sulfur (average) Tunnelton, 2.0% sulfur (average) (all mines located in west-central Pennsylvania)
Coal Production at Subsidiary Mines (1979):	Greenwich, 1,988,000 tons Rushton, 519,000 tons Tunnelton, 441,000 tons Total, 2,948,000 tons
Estimated Recoverable Reserves (January, 1980):	Operating mines, 80,099,000 tons Undeveloped reserves, 337,050,000 tons Total, 417,149,000 tons
Total Number of Preparation Plants owned by Pennsylvania Power and Light Company:	Four
Total Electrical Capacity (1980):	6,546 Mw
Sources of Electricity:	Five company-owned coal stations; two jointly-owned coal stations (81% of power production)
	One company-owned oil station; eleven combustion-turbine and five diesel peak-load stations (17.4% of power production)

	Two company-owned hydro-stations; one jointly-owned hydro-station (1.6% of power production)
Peak Load (Winter, 1980):	4,945 Mw (Winter, 1980)
Reserve Capacity:	24%
Growth of Demand for Electricity:	3.3% increase from 1978 to 1979; slight decrease in usage from 1979 to 1980
Service Area:	10,000 square-mile area in 29 counties of central-eastern Pennsylvania
Number of Customers (1980):	999,500 (1980)
Revenues:	
Total Operating Revenue (1980):	$885,451,000
Fuel purchases (1979):	$635,778,000 (62% of total operating revenue)
Rates (1980):	Residential, 4.34¢/Kwh Commercial, 4.28¢/Kwh Industrial, 3.10¢/Kwh Average, 3.87¢/Kwh

The Greenwich Collieries Preparation Plant is one of four facilities owned by Pennsylvania Power and Light Company (PP&L) that cleans coal extracted from the company's three underground mines. This plant employs a new type of jig that removes about 50 percent of the sulfur and ash from the raw coal feed. PP&L relies on the Greenwich Collieries plant as part of its sulfur dioxide compliance strategy, since the plant produces cleaned coal that has a low enough sulfur content to be legally burned without the need for additional sulfur dioxide controls.

PP&L has a number of subsidiaries, including the Pennsylvania Mines Corporation (PMC), which mines and cleans coal for PP&L generating stations. Among U.S. utilities in 1979, PP&L (through PMC) mined the tenth largest quantity of coal.

PMC reports that it sells both the coal it mines and cleans and the coal it purchases on the open market to the parent company, PP&L, on a "zero-profit" basis. PMC describes itself as a captive subsidiary that is not intended to generate a profit for the parent company. PP&L believes that captive mines assure a supply of coal meeting its specifications. Thus coal cleaning has been an important component of PMC's operations, since it enables the subsidiary to provide a consistent fuel to the parent company.

PP&L's Coal Cleaning Policy

The percentage of cleaned coal burned in PP&L generating stations ranges from 38.4 percent to 100 percent. This is because PP&L does not require that coal be cleaned, but only that it meet the specifications set by the utility. V. R. Burkhart, PP&L's Manager of Mining Technology and Engineering, states that these specifications are set to (1) insure "continuously effective boiler performance," and (2) comply with government sulfur and particulate emission regulations.

The following table provides information on PP&L's coal-fired generating stations, including the percentage of fuel that is cleaned:

Station	Unit	Size of Unit (Mw)*	In-Service Date	Fuel Pr cleaned
Montour	1	765	1972	92.0
	2	750	1973	
Brunner	1	334	1961	60.6
Island	2	390	1965	
	3	740	1969	
Martins Creek	1,2	300 (units combined)	1954-56	100
Sunbury	1,2	152 (units	1949	
	3	103 combined)	1951	38.4+
	4	134	1953	
Holtwood	17	73	1954	NA

*as of 1979
†units 3 and 4 combined

At present, regulations of the Pennsylvania Department of Environmental Resources that apply to PP&L's coal-burning power stations generally limit the average sulfur content of the coal burned during a 30-day averaging period to less than 2.3 percent. To stay within this limit, PP&L can use low-sulfur raw coal and/or cleaned coal at its power plants without the need for flue gas desulfurization systems. According to PP&L, raw and cleaned coal obtained through market purchases have an average sulfur content of about 2.0 percent. However, cleaning is necessary to make most of the raw coal from company-owned mines, such as the Greenwich Collieries, satisfy sulfur dioxide emission requirements. At Greenwich, raw coal averages 1.5 to 2.5 percent sulfur before cleaning and 1.4 to 1.7 percent sulfur after cleaning.

The use of low-sulfur western coal as an alternative strategy for complying with sulfur dioxide regulations was ruled out because of its low heating value and the high cost of transporting it to PP&L generating stations.

GREENWICH COLLIERIES PREPARATION PLANT

The Greenwich Collieries Preparation Plant is designed to clean both coarse and fine raw coal. The plant is owned by PP&L through its subsidiary PMC, and it supplies primarily PP&L's Montour Generating Station. This preparation plant is the first in the United States to employ a "Batac jig" (see the chapter Methods of Coal Preparation), which provides for the precise control of fine-coal cleaning. The use of this jig enables the plant to remove approximately 40 percent of the inorganic pyritic sulfur from the raw coal and significantly reduce its ash content. PP&L relies on this preparation plant to provide cleaned coal that will meet the sulfur dioxide and particulate emission requirements at the Montour Generating Station.

Effectiveness of the Cleaning Plant

The following table compares raw and cleaned coal characteristics at the Greenwich plant:

Effectiveness of the Cleaning Plant			
Coal Characteristics	Raw Coal	Cleaned Coal	Decrease / Increase
Sulfur (%):	1.5 to 2.5	1.4 to 1.7	-20 (average)
Ash (%):	25 to 33	15.0 maximum	-40 to 55
Moisture (%):	4.0 to 5.0	7.0 maximum	+40 to 75
Btu per pound :	10,300	12,500 minimum	21% increase
Btu recovery in cleaned coal: 96.1%			

The rocks and other undesirable materials in the coal are removed by two rotary breakers; these were recently installed to replace the Jeffery/Baum jigs, which the company reported as being less efficient. The Greenwich plant uses Batac jigs, thermal dryers, and other equipment to clean and remove water from the coal. The plant can operate either one, or both, of the two cleaning circuits at a given time, depending on the demand for cleaned coal and the need for maintenance or repair of either circuit.

Sources of Raw Coal

The preparation plant gets its raw coal from the Greenwich Collieries No. 1 and No. 2 mines, which are located near the plant. Raw coal is transported from the two mines to the plant on a conveyor belt. The coal that is now mined and cleaned at Greenwich is medium-volatile bituminous coal from the lower Freeport seam. The Greenwich mine has more recoverable reserves than any other active PMC mine— 62,603,000 tons as of 1980.

Problems with the Preparation Plant

The major difficulty encountered during the start-up of the fine-coal cleaning plant involved operating personnel that were inadequately trained. According to V. R. Burkhart, significant problems arose after the fine-coal plant equipment was installed, including frequent outages resulting from operating errors, long periods of downtime due to equipment failures, unacceptable amounts of time lost in locating problem areas, and long repairtime periods due to lack of familiarity with the equipment on the part of the mechanics. Burkhart states that the problem was solved by engaging the services of the management-consulting firm Kurt Salmon Associates to develop an in-house "school" for operating personnel and mechanics. A simulated control panel was constructed to train plant operators to operate the plant efficiently and minimize equipment downtime, and to identify and locate mechanical problems quickly. Mechanics were taught repair and maintenance skills. The training program has been termed "a success" by the company; PP&L reports that the plant now operates with greater reliability, that labor, maintenance and repair costs have been lowered, and that the quality control of the cleaned coal has improved. Burkhart feels that the costs of the training program were recovered one year after the plant personnel completed training.

Wastes

Wastes are removed from the plant by belt conveyor to an 80-acre disposal area about a half mile away; there they are spread by truck and compacted to satisfy Pennsylvania Department of Environmental Resources regulations. This site will be filled within the next five to six years, and an additional 100-acre site will be developed to accommodate the wastes produced during the remaining life of the mine (projected at 30 years). PMC did not provide data on the cost of waste disposal at the Greenwich plant.

The waste-disposal area is compacted in order to encourage water runoff. According to PMC, this runoff is collected in diversion ditches and requires only settling before it is discharged into a stream. PMC reports that all the water which is released meets Pennsylvania water quality standards. In addition, surface drainage facilities have been extensively modified to achieve compliance with U.S. Office of Surface Mining requirements.

PMC claims to have encountered "few" regulatory difficulties in the operation of the Greenwich Collieries Preparation Plant. According to PMC, the plant has a closed water circuit and does not itself discharge any water into the surrounding environment when it is in operation. An acid-drainage water-treatment plant processes all water from the mine and waste-disposal sites that requires treatment.

Capital and Operating and Maintenance Costs

The capital cost of the Greenwich Collieries plant was $13,721,055, or $19,856,701 in 1979 dollars with an 8 percent inflation adjustment. (See cost table at the end of this profile.) The annual operating and maintenance costs at Greenwich were estimated by INFORM at $5,369,000 in 1980.

MONTOUR GENERATING STATION

The Montour Generating Station is a 1,515 Mw twin-boiler plant that burns pulverized coal. Although PP&L does not specifically require that the coal burned at Montour be cleaned, 92 percent of it was cleaned in 1980. The Montour station burned raw coal exclusively for three years until 1976, when it switched to cleaned coal in order to comply with Pennsylvania's particulate and sulfur dioxide regulations. The Montour power plant must meet Pennsylvania's SO_2 emission standard of 4.0 pounds of SO_2 per million Btu. Coal is not blended at the plant unless supplied coal does not meet sulfur or ash specifications.

The PMC/PP&L-owned Greenwich Collieries Preparation Plant supplied the Montour power station with 75 percent of its coal needs in 1980. PP&L also purchased cleaned and raw coal for Montour from outside coal companies.

Coal Specifications

The following table lists the specifications set by PP&L for coal delivered to the Montour station:

Sulfur—2.2% maximum

Ash—15.0% maximum

Moisture—6.0% maximum

Btu per pound—12,400 minimum

Benefits and Problems of Burning Cleaned Coal

PP&L sees both savings and additional costs associated with its use of cleaned coal. Burkhart states that a major cost advantage for PP&L in burning cleaned coal is the fuel's low ash content, which cuts transportation costs. Also, low-ash cleaned coal is easier to grind than most raw coal, occasioning savings in the cost of pulverization; more cleaned coal than raw coal can be pulverized during a given time period, and pulverizer maintenance costs are therefore reduced. Boiler-ash disposal is cheaper when cleaned coal is burned, since less ash is present after combustion. In addition, low-ash coal generally causes fewer slagging problems.

Cleaned coal can either decrease or increase handling and storage costs. Although low-ash cleaned coal occupies less volume and is cheaper to store than raw coal, the higher moisture content of some cleaned coal may offset this advantage. Particulate control is improved when low-ash cleaned coal is burned, because there is less ash formed and thus the electrostatic precipitators (ESPs) are more efficient and remove more of the fine particulates. However, low-sulfur cleaned coal may interfere with the efficiency of the ESP, requiring ''conditioning'' of the flue gases with sulfur trioxide to improve particulate collection.

PP&L has not measured the savings realized by burning cleaned coal at the Montour station. The utility is therefore unable to determine whether the use of cleaned coal has resulted in higher or lower electricity costs to the consumer.

Transportation

Cleaned coal, ranging in size from two inches to 28 mesh, is shipped daily from the Greenwich plant to the Montour station, 196 miles away, by PP&L-owned unit trains. The cost of shipping the coal averages $3.60 per ton of cleaned coal.

Coal Costs

PP&L's costs for all cleaned coal delivered to the Montour station averaged $32.63 per ton (130¢ per million Btu) in 1979 and $35.65 per ton (143¢ per million Btu) in 1980. The utility did not provide exact figures for either Greenwich Collieries coal or coal bought from outside suppliers.

FUTURE PROSPECTS

In general, Burkhart perceives an increasing role for coal cleaning in the future of electricity generation, although this technology cannot produce a coal that can meet the newest federal New Source Performance Standards. He says, ''As transportation costs increase and coal seams become lower in quality with more impurities (including higher sulfur content) present in remaining reserves, coal cleaning will be an increasingly viable means of improving ROM [run-of-mine] coal for utility use.'' Burkhart hopes that when the Clean Air Act is reviewed in 1981, the sulfur dioxide reduction requirements will be ''tempered,'' making compliance with this legislation ''more practical'' for utilities. At the present time, he sees few government incentives to encourage either conventional coal cleaning or research into new technologies by utilities. He states, ''Unfortunately, many policymakers know too little about coal cleaning and realistically attainable results.''

GREENWICH COLLIERIES PREPARATION PLANT

Beginning of Commercial
Operation:

    Coarse-coal washing plant: Early 1971

    Fine-coal washing plant: January 1976

Construction Time:

    Coarse-coal washing plant: Slightly over one year

    Fine-coal washing plant: Two years, eight months

Architectural, Engineering and
Construction Firm:

    Coarse-coal washing plant: Lively Manufacturing and Equipment Company (a subsidiary of Elgin National Industries)

    Fine-coal washing plant: Roberts & Schaefer Company

Total Tons Processed

    Design capacity: 1,050 tons of cleaned coal per hour; 3,234,000 tons per year

    Operating capacity: 900 tons of cleaned coal per hour; 2,376,000 tons per year. Normally operates 14 hours per day, five days per week, 50 weeks per year

User: Pennsylvania Power and Light Company, Montour Generating Station

Manpower Required for
Plant Operation: Seven to ten persons per operating shift

Efficiency of Plant Operations

    Preventive maintenance
    requirements: 40 hours per week--one eight-hour maintenance shift per day

| Average downtime: | Not available. Problems frequently occur with thermal dryer controls, as well as miscellaneous plant-control, instrumentation, equipment and operating problems |

Resources

Energy rating:	5 Mw (6 Mw when loading unit train)
Water use:	Closed water loop --7 to 14 million gallons per day (10 gallons per ton of coal)
Magnetite requirements:	No magnetite used

| Wastes Generated: | 3,600 tons per day; 50% coarse, 50% fine |

COSTS

| Capital Cost of Plant: | Original cost, $13,721,055.09; estimated cost in 1979 with 8% increase per year, $19,856,701.15 |

| Annual Operating and Maintenance Costs: | $3,646,000 for the first eight months of 1980.
Estimated annual costs (1980), $5,369,000;
cost per ton of clean coal (1980), $2.75 |

Cost of Resources:

Energy:	$1,126,949 for electric power (1979)
Water:	Minimal (only pumping costs)
Magnetite:	No magnetite used

| Waste Disposal Costs: | NA |

| Price Charged Per Ton of Cleaned Coal: | NA |

| Transportation Costs Per Ton of Coal: | $3.60 per ton of cleaned coal |

| Financing Source for Plant: | Construction funded by leases (30%) and long-term borrowing from banks (70%), amortized over 20 to 25 years |

TENNESSEE VALLEY AUTHORITY
Chattanooga, TN 37401

Coal Cleaning Plant Profiled: Paradise Preparation Plant
Drakesboro, Kentucky

Mines Served: Ayrgem Mine (AMAX Coal Company)
Central City, Kentucky

Pyro (Pyro Mining Company)
Sturgis, Kentucky

Sinclair Mine (Peabody Coal Company)
Drakesboro, Kentucky

Power Plants Profiled: Colbert Generating Station*
Mussel Shoals, Alabama

Johnsonville Generating Station+
Johnsonville, Illinois

Paradise Generating Station**
Drakesboro, Kentucky

Coal Use (1980):   37,310,507 tons

Sources (1979):

Captive Mines: 22%
Long-Term Contracts: 74%
Spot Market: 4%

Coal Production at Ayrgem,
Pyro and Sinclair Mines (which
supply coal to TVA-owned
Paradise Preparation Plant):

Mine	Tonnage
Ayrgem:	1,913,320
Pyro:	1,800,000
Sinclair:	3,693,341

Estimated Reserves Owned
or Controlled by TVA: 730,000,000 Tons

Total Number of TVA-Owned
Preparation Plants: Four

Total Electrical Capacity (1980):   29,865 Mw

Sources: 12 TVA-owned coal stations
(57% of power production)

Two TVA-owned nuclear stations
(15% of power production)

29 TVA-owned hydro stations
(11% of power production)

Four TVA-owned combustion
turbine plants
(8% of power production)

One TVA-owned pumped storage
station
(5% of power production)

Eight U.S. Army Corps of
Engineers' Dams
(3% of power production)

12 Alcoa Dams
(1% of power production)

Peak Load (Winter 1980):   20,745 Mw

Reserve Capacity: NA

Growth of Electrical Demand: 2.6% increase between 1979 & 1980

Service Area: 80,000 square miles covering almost
                                    all of Tennessee, portions of Northern
                                    Alabama, Northern Mississippi, South-
                                    western Kentucky, and small parts of
                                    Georgia, North Carolina and Virginia.

Estimated Population: Seven million

Number of Customers   2,784,675
(September 1980):

Regulatory Commission: Federal Energy Regulatory Commission

Revenues:

Total Operating Revenue
(1980): $3,204,280,000
Fuel Purchases (1980): $1,301,221,000 (41% of total oper-
                                                   ating revenue)

Rates (1980):

Residential: 3.29¢/Kwh
Commercial: 2.71¢/Kwh
Federal Agencies: 2.54¢/Kwh
Municipal and Cooperative
Distributors: 2.56¢/Kwh

*Colbert burns cleaned coal from R&F Coal Company's preparation
plant in Warnock, Ohio.

†Johnsonville burns cleaned coal from Island Creek Coal Company's
No. 1 Mine preparation plant in Morganfield, Kentucky.

**Paradise receives its cleaned coal from the company-owned Paradise
preparation plant in Drakesboro, Kentucky.

The Tennessee Valley Authority (TVA) is one of the nation's largest utilities and a major user of cleaned coal. In 1980, TVA purchased 37.3 million tons of coal for its 12 coal-fired plants. This represented about 10 percent of the tonnage burned in domestic power production that year. At present, 60 percent of the coal burned in TVA's boilers is cleaned.

Two of TVA's preparation plants completed in 1981—its third and fourth—process among the highest tonnages of raw coal per hour in the nation. One of these, the Paradise plant, located near Drakesboro, Kentucky, is one of the largest U.S. facilities to use heavy-media equipment. It recently began shipping cleaned coal to TVA's Paradise generating station (June 1981). The Paradise station (profiled below) will be the only TVA power plant to use both cleaned coal and flue gas desulfurization (FGD) to meet sulfur dioxide emission standards once the FGD system is completed in 1984.

TVA, a wholly owned agency of the United States government, was established by an act of Congress in 1933 to develop the Tennessee River system and other resources of the Tennessee Valley.* Coal is now called "the solid backbone of TVA," having fueled 57 percent of its power production in 1980. However, TVA's power program was born with hydroelectric energy, which still supplies 11 percent of its power.

TVA's Coal Cleaning Policy

Coal cleaning is now a part of TVA's strategy for satisfying environmental regulations. For the present, TVA's 12 coal-fired power plants (63 units with an installed capacity of 17,796 Mw) are meeting sulfur dioxide emission standards with a combination of cleaned coal, FGD, electrostatic precipitators (ESPs), and baghouses.†

TVA generally sets sulfur specifications for its coal rather than stipulating the use of cleaned coal. Therefore the amount of cleaned coal burned at TVA plants varies, as described in the following chart.

---

*In addition to generating and selling electrical power, TVA is engaged in navigation, flood control, fertilizer research, and the conservation of natural resources through improvements in agriculture and forestry.

†TVA representatives claim that sulfur dioxide standards are indirectly dictating the use of cleaned coal or the installation of FGD systems.

Station	Number of Units	Total Capacity	In-Service Date	Fuel Pre-cleaned(%)*
Watts Bar	4	240	1942-45	95.3
Johnsonville	10	1,485	1951-59	36.4
Widows Creek	8	1,978	1952-65	43.7
Kingston	9	1,723	1954-55	56.7
Colbert	5	1,420	1955-65	100.0
Shawnee	10	1,750	1953-56	45.9
Gallatin	4	1,255	1956-59	100.0
John Sevier	4	847	1955-57	47.4
Bull Run	1	950	1967	0.0
Paradise	3	2,558	1963-69	0.0
Allen	3	990	1959	89.1
Cumberland	2	2,600	1972-73	48.3
	63	17,796		46.7 (weighted average)

*or blended with coal

Almost all TVA's coal, cleaned and uncleaned, is purchased by contract (approximately 74 percent), although the utility has mined and is now mining its own reserves. At present, TVA employs several "outside" coal companies to mine and clean its coal.*

TVA elected not to burn low-sulfur western coal at its power plants for two reasons: First, the Btu or heating value of western coal is too low to meet TVA's boiler specifications for Btu content. Second, since western coal is higher in ash as well as lower in Btus than most midwestern and eastern coals, and therefore has more bulk for a given amount of energy, more western coal would have to be fed through the pulverization equipment. This would necessitate increased pulverizer capacity, an added expense at the plant.

Research and Development

TVA's coal cleaning research and development program is unique among utilities. Two percent of the TVA Power Program's annual revenue (which was $2.657 billion in 1979) is used for R&D, and one fifth of this 2 percent is contributed to the Electric Power Research Institute. Additional funds for the Power Program's R&D come from the

---

*For example, Camp Breckinridge, a TVA-owned mine and preparation plant operated by Peabody Coal Company, supplies cleaned coal to TVA's Cumberland power plant.

U.S. Department of Energy or other federal agencies. The Power Program's Energy Demonstration and Technology Division, which covers all R&D of coal cleaning and FGD, has an annual budget of $35 million.

As a federal agency, TVA is compelled to share its R&D results. G.G. McGlamery, a TVA spokesman, points out:

> Anyone has direct access to our results; we publish positive and negative results on technical through economic assessments. The only problem this presents is in working with private industry. TVA can rarely sign secrecy agreements because most TVA projects are highly visible, and industry will generally not go along with this.

An R&D project now under way is examining the costs and benefits of burning cleaned versus raw coal at the Paradise Generating Station. This plant burned raw coal for 15 years and in June 1981 switched to burning cleaned coal.

However, "this is not the ideal test," notes Randy Cole, Project Manager, Combustion Systems Coal-Cleaning. He adds that "running identical boilers side by side, one with raw coal, the other with clean coal, would make it more like a [controlled] laboratory test." Since the Paradise boiler is 15 years older than when it began burning raw coal, and many variables have altered the physical structure and operating conditions of the boiler since that time, the exact savings attributable to cleaned coal cannot be calculated in this test, Cole reports.

PARADISE PREPARATION PLANT

As noted earlier, TVA's Paradise Preparation Plant is one of the largest plants in the United States to use heavy-media cleaning equipment; it produces 1,607 tons of cleaned coal per hour. TVA may test High Gradient Magnetic Separation (HGMS), an advanced physical coal cleaning technology (see Oak Ridge National Laboratory profile), in its fine-coal circuit at Paradise in 1983. HGMS may save TVA money, since its use makes costly water-removal equipment unnecessary. Paradise was the second most expensive preparation plant in INFORM's sample of 12, costing $44 million in capital expenses, plus

$68 million for conveyor and "auxiliary" equipment, in 1981 dollars. According to TVA, the Paradise Preparation Plant enables it to continue using local high-sulfur coal. If burned without prior cleaning, this coal would emit more than 5.2 lb $SO_2/10^6$ Btu, the Kentucky sulfur dioxide emission ceiling.

Because its raw coal was causing slagging and fouling problems at the Paradise Generating Station, TVA originally decided to build the preparation plant in order to reduce the coal's ash content. The utility later chose to develop a more sophisticated (level 4) preparation plant at Paradise, to enable its existing coal supplies to meet Kentucky sulfur dioxide emission standards,* as well as to reduce ash.

Effectiveness of the Cleaning Plant

The following table compares the characteristics of raw and cleaned coal at the Paradise Preparation Plant (data as of March 1981):

Coal Characteristics	Raw Coal	Cleaned Coal	Decrease / Increase
Sulfur (%):	4.90	3.10	-36.7
Ash (%) :	20.10	9.00	-55.2
Moisture (%):	10.70	12.20	+14.0
Btu per pound :	11,407	13,206*	+15.8%

Btu recovery in cleaned coal: 93%

*When burned, raw coal emits 8.2 lb $SO_2/10^6$ Btu; cleaned coal emits 5.0 lb $SO_2/10^6$ Btu.

The Btu value of the preparation plant wastes has been calculated by TVA to be 6,970.

The Paradise Preparation Plant consists of four coal cleaning circuits, each housed in a separate building, that can operate simultaneously or independently. This design is unique among INFORM's sample of 12 plants, none of which contain more than two cleaning circuits. The use of four circuits allows TVA to vary production rates and maintenance schedules with great flexibility.

Although thermal dryers were originally part of the Paradise

---

*Raw coal emits 8.2 lb $SO_2/10^6$ Btu (when burned); cleaned coal emits 5.0 lb $SO_2/10^6$ Btu.

plant's design, and buildings were constructed to hold them,* they have not been installed. TVA has found that vacuum filters and other dewatering equipment "can do an adequate job" of removing water from coal and wastes.

Four basic types of cleaning equipment are used at the Paradise plant: heavy-media washers to clean larger-size coal (more than 3/8 inch), heavy-media cyclones for medium-size coals (3/8 inch to 28 mesh), and hydrocyclones and froth-flotation units to clean the finest-size coal particles (less than 28 mesh).

Sources of Raw Coal

TVA obtains raw coal for the Paradise Preparation Plant from the Ayrgem, Pyro and Sinclair mines located in nearby Kentucky counties. The plant burns Muhlenburg County 9, 11, and 12 seams, western Kentucky, Type-B bituminous coal. Coal is hauled from the mines to the preparation plant by 100 ton off-road trucks and by conveyor belt.

Preparation Plant Problems

Because the Paradise plant only recently started operations (June 1981), TVA has not yet reported any problems with plant equipment and/or cleaning efficiency.

Wastes

Each type of equipment at the Paradise Preparation Plant produces a different quantity of waste. The coarse-coal equipment produces 111 tons of wastes per hour, the medium-coal equipment, 148 tons per hour, and the fine-coal equipment, 34 tons per hour. Assuming a 16-hour production schedule, 4,825 tons of wastes are generated daily.

TVA has purchased 2,600 acres of land adjacent to the plant to be used for waste disposal (by landfill) during the projected life of the plant — 20 to 30 years. Wastes are moved to the landfill site by conveyor or

---

*One thermal dryer was to be shared by two coal cleaning circuits.

truck. All acid drainage from coal and waste piles, and from the preparation plant itself, is collected, treated and stored in a pond before reuse.

TVA reports that it has not yet been able to compute waste-disposal costs.

Capital and Operating and Maintenance Costs

The Paradise plant was second only to Pennsylvania Electric Company's Homer City Preparation Plant in capital cost. (It should be noted that a sizeable percentage of Homer City's capital cost was for rebuilding the plant, which has encountered extensive difficulties; see Pennsylvania Electric Company profile.) The Paradise plant itself cost $44 million; conveyors and auxiliary equipment, $68 million more. TVA reports operating and maintenance costs of $1.39 per ton of cleaned coal.

COLBERT, JOHNSONVILLE AND PARADISE GENERATING STATIONS

The TVA-owned Colbert (5 units, 1,420 Mw), Johnsonville (10 units, 1,485 Mw) and Paradise (3 units, 2,558 Mw) generating stations all burn cleaned coal in order to comply with sulfur dioxide emission regulations and reduce ash to improve boiler performance.

These stations were chosen for INFORM's study because the cleaned coal they use is supplied by preparation plants profiled in this report (see also R&F Coal Company and Island Creek Coal Company profiles).

Colbert receives cleaned coal from R&F Coal Company's Warnock, Ohio, preparation plant; Johnsonville receives cleaned coal from Island Creek Coal Company's Hamilton No. 1 Mine preparation plant. Paradise recently began receiving cleaned coal (which it is now stockpiling and plans to begin burning in December 1981) from TVA's own Paradise preparation plant.

Six citizens' suits were filed in the middle 1970s under the Clean Air Act alleging that sulfur dioxide emissions from ten of TVA's coal-fired power plants, and particulate emissions from seven plants, violated standards set by individual state implementation plans. In December 1978, a proposed settlement specified compliance schedules for controlling sulfur dioxide and particulates, and provided for stipulated daily penalties if TVA did not satisfy emission standards. To meet the schedule for particulates, TVA added new ESPs; to meet the schedule for sulfur dioxide and at the same time continue to use coal supplied under existing contracts, TVA built the Paradise Preparation Plant (to reduce its coal's sulfur content) and will install scrubbers on Paradise Generating Station Units 1 and 2.

Following are the sulfur dioxide emission standards with which the three TVA power plants must comply:

Plant	SO_2 Emission Standard
Colbert	4.0 lb $SO_2/10^6$ Btu
Johnsonville	4.5 lb $SO_2/10^6$ Btu*
Paradise	5.2 lb $SO_2/10^6$ Btu

*By 1983 the Johnsonville plant will have to meet a 3.4 lb $SO_2/10^6$ Btu emission standard

As mentioned above, TVA usually specifies the quality of the coal it purchases. Coal does not have to be cleaned to meet TVA's specifications.

Coal Specifications

The specifications set for coal, either cleaned or raw, provided by captive and outside coal suppliers are the following:

	Colbert	Johnsonville	Paradise
Sulfur (%):	2.20	2.30	4.20
Ash (%):	12.10	9.80	16.90
Moisture (%):	7.80	10.30	9.10
Btu per pound:	11,680	11,600	10,570

Benefits of Burning Cleaned Coal

TVA sees several advantages of burning cleaned coal as opposed to raw coal. First, its use has enabled the utility to burn more locally mined western Kentucky coal (and still meet government pollution-control standards) than it could if only raw local coals were available. In fact, TVA chose to build a more sophisticated version of the Paradise Preparation Plant than it had originally intended, in order to meet state sulfur dioxide emission ceilings with local coal.

Second, as TVA points out, because of the industrywide problems resulting from lower-quality coal being mined (because of new mining practices and the declining availability of high-quality seams), coal cleaning at TVA has led to three beneficial effects on power generation: (1) improved availability of boilers, and therefore fewer purchases of "outside" electric power; (2) reduced maintenance costs at generating stations, due mostly to shorter plant outages; and (3) overall "smoother operations."*

Third, TVA finds three benefits from the use of cleaned coal in combination with FGD at utility power plants: 1) lower capital costs for scrubbers, 2) increased availability of both scrubbers and boilers, and 3) lower sulfur dioxide removal rates required of the scrubber.

TVA notes several more advantages of cleaned coal. Transportation costs can be reduced when a fuel with less waste is hauled. Centralized coal cleaning plants could provide fuel with a more consistent quality, and this might encourage the purchase of coal from smaller mining operations that could not afford to build and operate preparation plants themselves.

Problems of Burning Cleaned Coal

According to TVA, the cost of coal cleaning is its major disadvantage. Lesser drawbacks are related to the difficulties in coal feeding often caused by the higher moisture in cleaned than in raw coal. Finally,

---

*TVA reports that it had severe fly ash problems when high-sulfur coal was burned at Paradise; difficulties were "premature breakdowns" and "excessive maintenance." As described above, TVA planned to build a coal cleaning plant even before environmental (SO_2) problems encouraged it to clean its coal more extensively.

low-sulfur cleaned coal may require more extensive and more energy-intensive equipment (e.g., larger ESPs) to control fly ash emissions.

Transportation

TVA uses barges or conveyors to ship cleaned coal from its sources to the power stations. In the case of the Paradise station, a conveyor belt moves the cleaned coal from the preparation plant to a storage pile or to the power plant for less than $0.35 per ton. For the Colbert plant, 50,000 tons of cleaned coal per week are shipped 720 miles from the R&F Preparation Plant on 33 barges. The barges are unloaded daily and make one tow each week at a cost of $10.97 per ton. TVA hauls 34,000 tons of coal per week by barge from the Hamilton No. 1 Mine preparation plant to the Johnsonville generating station, 210 miles away. Twenty-three barges complete one tow per week. Transportation cost per ton are $2.29.

Coal Costs

TVA did not provide figures on the price per ton of the cleaned coal burned at its power plants. Robert L. Frank, TVA's Project Manager of Coal Cleaning Performance & Reliability, termed cleaned coal contracts "more expensive than raw coal contracts," but noted that it was "very easy" to obtain clean coal contracts "if you [are willing to] pay the price."

Raw coal delivered to the Paradise Preparation Plant is $22.64 (May 1981); it may rise to $24 per ton by 1985. Frank noted that the cost of cleaned coal is calculated by adding the raw coal cost to the preparation plant's operating costs and loan amortization.

FUTURE PROSPECTS

According to Frank, eventually most coal will be cleaned, whether it is to be burned directly, liquified, or gasified. Thus he forecasts an increased demand for cleaned coal. He views utilities as future investors in preparation plants, primarily because they need to comply with sulfur dioxide regulations at their power stations and because cleaned coal can offer increased boiler availability. Frank believes that most coal will be cleaned to level 4 (coarse, medium, and fine cleaning; see Methods of Coal Preparation). His company has a stake in High Gradient Magnetic Separation, which may be incorporated in the Paradise Preparation Plant for testing in 1983. Frank observes, "economics will dictate its use."

PARADISE PREPARATION PLANT

Beginning of Commercial Operation:	June 1981
Construction Time:	Two years, 11 months
Architectural, Engineering and Construction Firm:	Roberts & Schaefer Company
Total Tons Processed:	1,607 tons of cleaned coal per hour
	Manned to operate 16 hours per day, five days per week, 52 weeks per year
	(In March 1981, 25,500 tons of raw coal were processed, 18,800 tons of clean coal were produced)
Users:	All cleaned coal used by TVA to generate electricity
Manpower Required for Plant Operation:	NA
Efficiency of Plant Operations	
Preventive maintenance requirements:	Eight hours per day
Average downtime:	NA
Resources	
Energy rating:	10 Mw capacity; 2 Mw to 3 Mw is plant's normal load
Water use:	195 Gallons per minute added as make-up water
Magnetite requirements:	Approximately 320 tons of magnetite per month (or 2 pounds per ton of raw coal) in heavy-media equipment*
Wastes Generated:	4,528 tons per day (16-hour production schedule)

*TVA reports that the plant will use 838 pounds per hour in its heavy-media vessels and 936 pounds per hour in its heavy-media cyclones, for a total of 1,774 pounds per hour.

COSTS:

Capital Cost of Plant (1981 dollars):	$44 million--preparation plant
	$68 million--conveyors and auxilliary equipment
Annual Operating and Maintenance Costs:	$1.39 per ton of cleaned coal

Cost of Resources

Energy:	NA
Water:	NA
Magnetite:	NA
Reagents:	$0.15 per ton of raw coal
Waste Disposal Costs:	NA
Price Charged Per Ton of Cleaned Coal:	TVA claims that it is "too early to tell" how much coal cleaning costs
Transportation Costs Per Ton of Coal:	$0.35 per ton or less
Financing Source For Plant:	Government bond issues; revenues from the sale of electric power

Utilities Burning Purchased Cleaned Coal

ASSOCIATED ELECTRIC COOPERATIVE, INC.
2814 South Golden
Springfield, MO 65801

Power Plant(s) Profiled:	New Madrid Station Units Nos. 1 & 2 Marston, MO
	Thomas Hill Station Units Nos. 1, 2 & 3 Moberly, MO

Coal Use (1979):	3,970,512 tons
Total Electrical Capacity (1980):	1,728 Mw
Sources:	100% from coal-fired stations: Thomas Hill, Units Nos. 1 & 2 New Madrid, Units Nos 1 & 2
Peak Load (1980):	1,856 Mw
Reserve Capacity:	NA
Growth of Electrical Demand (1979-80):	6.0%
Service Area:	1,661 miles of up to 345 Kv transmission lines scattered across the rurals of Missouri and Southeastern Iowa

Number of Members:	43 distribution cooperatives who deliver AECI's electricity to 400,000 members in rural Missouri and South-eastern Iowa. On the average, these distribution cooperatives provide electricity to 6 to 9 customers per square mile (far less than the 40 to 60 customers per square mile of a typical non-cooperative utility)
Regulatory Commission:	Rural Electrification Administration
Revenues:	
Total Operating Revenue (1980):	$206,097,000
Fuel Purchases:	44% of 1980 expenditures
Rates:	NA

Associated Electric Cooperative, Inc. (AECI), was created in 1962 by six generation and transmission cooperatives (G&Ts) across Missouri to plan, construct and operate their generation and transmission facilities.* In forming AECI, the G&T cooperatives sought to pool their resources to build larger, more efficient power stations.

By 1982 all the coal burned at AECI's two coal-fired power plants will be cleaned. AECI burns cleaned coal to improve plant operations at its New Madrid Station and to meet environmental regulations at the Thomas Hill Station. The cooperative is now building a preparation plant at an AECI-owned mine to supply most of its coal.

Coal-Fired Power Plants

·Listed below are the number and type of generating units, date of installation, and megawatt capacity for the coal-burning plants, New Madrid and Thomas Hill. To handle projected future demand, AECI plans to open a third coal-fired unit at the Thomas Hill plant in 1982.†

---

*Through these six G&Ts, AECI is indirectly owned by more than 400,000 rural electric cooperative members.

†AECI claims that rising fuel costs, high interest rates, and increasing government regulation will make the new unit cost $460 million, or $730 per net Kw of capacity. In contrast, the two original Thomas Hill units, completed in 1966 and 1969, each cost $66 million ($137 per Kw).

Station and Unit	Installed	Type of Unit	Capability	Fuel Pre-cleaned(%)
New Madrid				
Unit 1	1972	Cyclone-fired, wet bottom	600	100
Unit 2	1977	Cyclone-fired, wet bottom	600	100
Thomas Hill				
Unit 1	1966	Cyclone-fired, wet bottom	180	[Blend of raw and cleaned coal]
Unit 2	1969	Cyclone-fired, wet bottom	180	"
Unit 3	(1982)	Fired by pulverized coal	(670)	"

According to AECI, burning cleaned coal will provide increased electricity by decreasing boiler downtime at New Madrid and help the cooperative comply with legislation governing air-pollutant emissions for those boilers using Missouri coal. At present, the New Madrid Station burns cleaned coal only, and Thomas Hill Units 1 and 2 use a blend of raw and cleaned coal.

AECI is now building a preparation plant that uses heavy-media equipment at its Prairie Hill Mine. When the plant is completed, it will provide all of the coal for Thomas Hill Units 1 and 2. Unit 3 will either burn 100 percent cleaned coal or a blend of raw and cleaned coal. The cleaned coal burned in this unit will also come from AECI's new preparation plant.

NEW MADRID STATION, UNITS 1 AND 2

Units 1 and 2 of the New Madrid Station, located in the Bootheel area of Missouri, were completed in 1972 and 1977, respectively. Together they have the capacity to produce 1,200 Mw of electrical power. The two units must meet a sulfur dioxide emission standard of 10.0 pounds of SO_2 per million Btu (10.0 lb $SO_2/10^6$ Btu), which is a much less stringent standard than most of those reported by utilities in INFORM's sample.

Sources of Coal and Coal Specifications

The New Madrid Station was originally designed to burn raw coal from Peabody Coal Company mines located in southwestern Illinois. However, because of the large percentage of fire-clay in this coal, the boiler tubes in Units 1 and 2 became partially plugged when the coal was burned. As a result, the amount of electricity which the units could generate was reduced, occasionally forcing AECI to purchase higher-priced emergency power from other utilities. In response to this problem, the New Madrid Station began burning cleaned coal.

In 1979 the Peabody Coal Company supplied New Madrid with 90.23 percent of its coal. This coal was cleaned at Peabody's Randolph and River King preparation plants in Illinois before being shipped. (For information on the Randolph Preparation Plant, see Peabody profile.) The additional coal for the New Madrid Station came from spot-market purchases. In 1980, the New Madrid Station began burning 100 percent cleaned coal, although according to AECI's coal-supply agreement, deliveries of 100 percent cleaned coal were not required until 1983.

The specifications set by AECI to be met by coal suppliers for Units 1 and 2 of New Madrid, and the average sulfur, ash, moisture and Btu content of the fuel received, are:

	Specifications Set	Contents of Coal Received 1979	1980
Sulfur (%):	4.0 maximum	NA	3.19
Ash (%):	12.0 maximum	11.8	9.90
Moisture (%):	15.0 maximum	13.9	13.40
Btu per pound:	10,500 minimum	10,584	10,827

Benefits and Problems of Burning Cleaned Coal

Boiler availability and generating capacity at New Madrid have increased since it began burning 100 percent cleaned coal in 1980, and AECI has been able to reduce the consumption of expensive emergency power. However, AECI has not recorded the increase in kilowatts generated by Units 1 and 2 when they are burning cleaned as opposed to raw coal.

AECI has also noted positive effects of cleaned coal on coal handling and storage, ash disposal and particulate removal at New Madrid. By reducing the amount of ash in the coal, cleaning decreases the amount of coal that must be handled and stored, the amount of ash left in the boiler after combustion that must be disposed of, and the amount of particulates that must be removed. Because New Madrid's boilers have cyclone burners which do not need pulverized coal, coal cleaning does not benefit the operation of pulverizers.

AECI noted that fugitive dust emissions have remained unchanged since the substitution of cleaned coal for raw coal. The cooperative continues to use only the dust-control measures it used when it burned raw coal: a baghouse dust-suppression system in the crusher house and a dust-ventilation system in the "bunker" area.

Transportation

New Madrid is 220 miles from the Marissa underground mining complex, which is served by the Randolph Preparation Plant, and 280 miles from the River King surface mine, which is served by the River King Preparation Plant. Each week about 48 barges bring approximately 70,000 tons of cleaned coal to the New Madrid plant from these two mines. In 1980 the cost per ton of shipping coal to New Madrid from the Randolph plant averaged $2.26.

Coal Costs

The average cost of cleaned coal for AECI's New Madrid Station in 1980 was $26.60 per ton, or $1.23 per million Btu. In 1979 the cost was $23.47 per ton, or $1.11 per million Btu.

AECI did not quantify the savings attributable to the burning of Peabody's cleaned coal versus raw coal at the New Madrid Station. However, when asked how changing to cleaned coal affected its annual fuel costs, AECI responded: "Overall, slightly increased costs for cleaned coal are offset by more efficient boiler operation." AECI also noted that ash-disposal costs have declined.

THOMAS HILL STATION, UNITS 1, 2 and 3

By 1982, when Unit 3 is completed, the Thomas Hill Station will generate over 1,100 Mw of electricity. The generating capacities of the existing and planned units at Thomas Hill are shown in the table at the beginning of this profile. At the present time, Thomas Hill Units 1 and 2 must meet a sulfur dioxide standard of 9.5 pounds of SO_2 per million Btu (9.5 lb $SO_2/10^6$ Btu). In 1982, when Unit 3 begins operation, the sulfur dioxide standards for Units 1 and 2 will drop to 8.0 lb $SO_2/10^6$ Btu. Unit 3 will have to meet a 1.2 lb $SO_2/10^6$ Btu New Source Performance Standard (NSPS).

Sources of Coal

On January 1, 1981, AECI entered the coal mining industry by buying Peabody Coal Company's Prairie Hill and Bee Veer surface mines. At present AECI is mining only the 20,000-acre Prairie Hill reserves, which are expected to provide 3.5 million tons of coal per year for the next 30 to 35 years. Coal from the Bee Veer mines is of low quality and AECI does not expect to mine it. The Thomas Hill Station will consume all the coal produced by the Prairie Hill mine.

Thomas Hill Units 1 and 2 now burn a blend of raw and cleaned coal to meet their sulfur dioxide standard (9.5 lb $SO_2/10^6$ Btu). Most of the coal used at the Thomas Hill power plant comes from the nearby Prairie Hill mine and a small amount is purchased from an outside supplier. Part of the Prairie Hill coal is cleaned in AECI's small (400 ton per hour) Bee Veer jig washing plant. To improve coal quality further and to insure compliance with the sulfur dioxide standard for Thomas Hill Units 1 and 2, AECI now employs selective-mining techniques at the Prairie Hill mine. Selective mining has decreased the coal's ash content and as a result, increased the coal's Btu content by 500 Btu's per pound. However, the use of selective-mining techniques has caused production to fall; when the new Thomas Hill Unit 3 starts operating, AECI will have to stop selective mining in order to increase coal production by one third (to provide fuel for this unit in addition to Units 1 and 2).

In the future, coal from Prairie Hill will be cleaned at a new and larger preparation plant being constructed by Roberts & Schaefer for AECI outside the Thomas Hill gates at the Prairie Hill mine face. This plant will clean the coal more extensively than does the Bee Veer facility. Designed to process 1,500 tons of raw coal per hour, it will use heavy-media cleaning equipment to increase the Btu content of the Prairie Hill coal from 8,800 Btu per pound to 10,800 Btu per pound and decrease the sulfur content from 11.55 lb $SO_2/10^6$ Btu to 7.18 lb $SO_2/10^6$Btu. This reduction in sulfur content will enable Thomas Hill Units 1 and 2 to meet their new sulfur dioxide standard without the need to blend Prairie Hill coal with lower-sulfur coal from other sources.

In order to meet the 1.2 lb $SO_2/10^6$ Btu emission standard, Thomas Hill Unit 3 will employ a combination of limestone scrubbers (flue gas desulfurization equipment) and either 100 percent cleaned coal or a mixture of raw and cleaned coal. Burning at least partially cleaned coal in Unit 3, in addition to operating scrubbers, will insure that the boiler consistently meets the NSPS. If Unit 3 burned only raw coal from Prairie Hill, which is high in sulfur, fluctuations in the sulfur content of the coal would cause the unit to exceed the sulfur dioxide emission limit at times. Cleaning Missouri coal before it is fed into the boiler and the scrubber, notes AECI, allows the utility to fulfill its commitment to "using Missouri coal and Missouri resources to generate the power needed by Missouri people."

Waste Disposal

All of the preparation plant wastes from AECI's new operation will be solid after coarse refuse goes through centrifuges and the fines go through static thickeners and vacuum filters. AECI plans to dispose of these wastes by depositing them in strip mine cuts. Eventually the mined land will be reclaimed. When Unit 3 is operating, sludge from the scrubber will be placed in the inactive portion of the Prairie Hill Mine.

Benefits of Burning Cleaned Coal

Although the cost-benefit trade-offs of coal cleaning cannot yet by computed for the Thomas Hill Station, AECI expects coal cleaning to decrease the limestone requirements for the Unit 3 scrubber and improve its efficiency. In addition, the utility believes that its ownership of coal reserves and a coal preparation plant will insure the reliability and quality of its coal supplies, decrease transportation costs, and lower the cost of electricity to its members.

CAROLINA POWER and LIGHT COMPANY
P.O. Box 1551
Raleigh, NC 27602

Power Plant Profiled:	Roxboro station, Units 1 to 4 Hyco, North Carolina
Coal Use:	9.2 million tons in 1980, mostly from mines in Kentucky, Virginia and West Virginia
Total Electrical Capacity (1980):	8,053 Mw
Sources of Electricity:	Coal, 56% of capacity Nuclear, 28% of capacity Hydro, 3% of capacity Oil, 13% of capacity
Peak Load (Winter 1980):	6,139 Mw
Reserve Capacity:	24%
Growth of Demand for Electricity:	3.9% from 1979 to 1980 (peak-to-peak increase); average annual consumption increases, 6.6%
Service Area:	Approximately 30,000 square miles in North and South Carolina; estimated population, 3 million
Number of Customers:	742,000
Regulatory Commissions:	North Carolina Utility Commission South Carolina Public Service Commission
Revenues:	
Total operating revenues (1980):	$1,075,604,000
Fuel purchases:	$411,191,000 (38% of total operating revenues)
Rates (1980):	Residential, 4.35¢/Kwh Commercial, 3.96¢/Kwh Industrial, 3.03¢/Kwh

Carolina Power & Light Company (CP&L) is one of the major coal users among utilities serving the Southeast. Like most utilities profiled by INFORM, CP&L does not specify cleaning for the coal it purchases; rather, it sets specifications for the sulfur, ash moisture and Btu content of the fuel as well as for other chemical constituents. About 50 percent of the 7.5 million tons of coal burned by CP&L in 1979 was either partially or fully cleaned.

The Roxboro generating station, profiled here, is a 2,355 Mw coal-fired facility located in Hyco, North Carolina. It is by far the largest of the utility's seven coal-fired plants, generating over four times as much power as any of the others in 1979 (9.6 million Kwh). In that year all the coal delivered to the Roxboro station was partially or fully cleaned and 96 percent was partially or fully cleaned in 1980. Twenty companies (31 mines) and five preparation plants supplied this fuel, including A.T. Massey's Marrowbone Development Corporation preparation plant in Naugatuck, West Virginia. (For information on A.T. Massey and the Marrowbone plant, see A.T. Massey profile.)

Coal-Fired Power Plants

The following table describes CP&L's six coal-burning plants of over 100 Mw:

Station & Unit	Size of Unit (Mw)	In-Service Date	Fuel Pre-cleaned(%)
Asheville	392		NA
Unit 1	(units	1964	
Unit 2	combined)	1971	
Cape Fear	316		NA
Unit 5	(units	1956	
Unit 6	combined)	1958	
H.F. Lee	252		NA
Unit 3		1962	
H.B. Robinson	174		NA
(burns coal and gas)			
Unit 1		1960	
Roxboro	2,355		96%
Unit 1	(units	1966	
Unit 2	combined)	1968	
Unit 3		1973	
Unit 4		1980	
L.V. Sutton	588		NA
(burns coal and oil)	(units		
Unit 1	combined)	1954	
Unit 2		1955	
Unit 3		1972	

The company expects to burn 258 million tons of coal in its existing and planned coal-fired units over their lifetimes. Forty million tons will be provided by a CP&L subsidiary, Robert Coal Company; 150 million tons, through existing long-term contracts; the balance (68 million tons), through additional long-term contracts, spot-market purchases, and possibly the acquisition and development of additional reserves.*

Coal Cleaning Policy

In order to comply with sulfur dioxide emission regulations, CP&L specifies the sulfur content of the fuel that it purchases. As long as the coal can meet these and other specifications, either cleaned or raw coal may be used. The average sulfur content of the coal bought by CP&L for its generating stations is quite low (less than 1.3 percent sulfur for a 12,000 Btu per pound coal). To satisfy the State Implementation Plan (SIP) in North Carolina, older boilers burning a fuel containing 12,000 Btu per pound must use coal with a sulfur content of 1.4 percent or less. South Carolina's SIP is satisfied by a 2.1 percent sulfur content.

Carolina Power and Light indicated that it spent $46.2 million in 1980 to conform to environmental regulations. About $18.2 million of this amount was used for air quality control and $24.7 million for water quality control. The company estimates that its 1981 figure will be $52 million.

CP&L does not believe it is realistic to bring low-sulfur western coal to the eastern states, because the transportation costs would be too high. In addition, CP&L's boilers are not designed to burn coal with a Btu content as low as that of most western coals (its boilers were intended for coal of no less than 12,000 Btu). The company notes also that some utilities have encountered slagging and fouling problems in their boilers when Appalachian and western coals were blended. Money is the critical concern in CP&L's decision not to burn western coal; if the company found it ultimately cheaper to burn these coals, it would.

---

*In 1974 CP&L entered into an agreement with Picklands Mather & Company, a firm engaged in owning, operating and managing properties, to develop two deep coal mines in Pike County, Kentucky. CP&L owns 80 percent of the two companies formed to develop the mines.

ROXBORO GENERATING STATION

The 2,355 Mw Roxboro station has four generating units. Three were built between 1966 and 1973, and the fourth started operating in the fall of 1980.

Units 1, 2 and 3 must meet an emission standard of 2.3 pounds of SO_2 per million Btu. Because construction of Unit 4 began after 1971 and before 1978, its SO_2 emissions must stay under the 1.2 lb $SO_2/10^6$ Btu stipulated in the federal New Source Performance Standards (NSPS).

Sources of Coal and Coal Specifications

Larry Yarger, Manager of CP&L's Fossil Fuel Section, who is responsible for acquiring the fuel for all the company's coal-fired plants, said that virtually 100 percent of the Roxboro station's coal needs are met by CP&L's captive reserves and by long-term contracts with coal suppliers.

To insure compliance with the NSPS at Unit 4, CP&L plans to furnish at least part of the unit's coal supply from its captive reserves in Pike County, Kentucky. This coal contains over 12,000 Btu per pound and has a very low sulfur content (less than 0.6 percent).

All coal supplied to the Roxboro Station must meet the following specifications set by CP&L:

Sulfur: 1.4% maximum (Units 1, 2 & 3)
         0.7% maximum (Unit 4)

Ash: 15% maximum

Moisture: 8% maximum

Btu per pound: 12,000 minimum

Benefits and Problems of Burning Cleaned Coal

Yarger sees quality consistency as the biggest advantage of using cleaned coal followed by ash reduction, Btu enhancement and in some cases, sulfur reduction. According to Yarger, coal cleaning is not used to reduce the sulfur content of coals purchased by CP&L because these coals are already low in sulfur.

All coal does not need cleaning, cautions Yarger; CP&L buys some high-quality coal that does not require washing. Coal cleaning, he states, should remain an option—offering a utility flexibility in meeting air quality standards—rather than another requirement for an industry already overburdened with regulations.

It is Yarger's opinion that most coal will need to be cleaned in the future, owing to modern mining techniques and the depletion of many of the country's high-quality coal seams. He notes that older boilers were designed for coal with a relatively high Btu content, and that such coal is not as available as it once was. In fact, says Yarger, CP&L's new units (Mayo 1 and 2) are being designed to burn coals meeting a broader range of specifications than the range met by coals burned in existing units, to account for poorer (lower Btu) coal quality. He observes that "now, some washed coals are not of as good quality as most raw coal was several years ago."

The use of cleaned coal is not thought to affect the operation of the electrostatic precipitators (ESPs) used for particulate control at all 18 of CP&L's coal-fired boilers. Although the company attributes operating problems with its ESPs to factors such as coal chemistry and its influence on ash resistivity, and boiler-tube leaks that contribute to unfavorable moisture conditions within the ESP, it sees no adverse role on the part of cleaned coal.

CP&L has not calculated the expenses or savings attributable to burning cleaned versus raw coal.

Coal Costs

CP&L would not provide figures for the cost per ton or per million Btu of the coal obtained from A.T. Massey's Marrowbone preparation plant or from any other supplier.

CENTRAL ILLINOIS LIGHT COMPANY
300 Liberty Street
Peoria, IL 61602

Power Plant Profiled:	E.D. Edwards Station, Unit No. 3 Peoria, Illinois

Coal Use (1980):	2,487,000 tons
Total Electrical Capacity (1980):	1,152 Mw
Sources:	Coal, 97.5% of capability Natural Gas, 2.5% of capability
Peak Load (1980):	1,091 Mw
Reserve Capacity (1980):	5.3%
Growth of Electrical Demand (1979-80):	3.4%
Service Area:	Electricity to more than 3,700 square miles and natural gas to a 4,500 square mile area in cen- tral Illinois
Number of Customers (1980):	241,049
Regulatory Commission:	Illinois Commerce Commission
Revenues:	
Total Operating Revenue (including gas sales)(1980):	$394,821,000
Electric Operating Revenue (1980):	$254,181,000
Fuel Purchases:	$94,391,000 (37% of Electrical Oper- ating Revenue)
Rates:	
Residential:	6.04¢ per Kwh
Commercial:	6.56¢ per Kwh
Industrial:	3.95¢ per Kwh
Other:	5.50¢ per Kwh
Average:	5.12¢ per Kwh

In order to meet environmental standards as economically as possible, Central Illinois Light Company (CILCO) decided to ship in low-sulfur western coal. CILCO developed a program that includes the use of both low and high-sulfur, raw and cleaned coals. CILCO's E.D. Edwards Unit 3, which is described in this profile, is one of the two boilers in INFORM's sample that burns cleaned low-sulfur western coal. It obtains this coal from the Western Slope Carbon Preparation Plant in Colorado. (See Northwest Coal Corporation profile.)

When CILCO considered the age of the equipment, space restrictions, problems of sludge disposal, and economic factors at each of its three coal-burning power plants, it elected to use cleaned low-sulfur coal at the Wallace and E.D. Edwards stations, and cleaned high-sulfur coal in combination with flue gas desulfurization (FGD) at the Duck Creek Station. Flue gas conditioning systems are used at the Wallace and E.D. Edwards stations to improve electrostatic precipitator performance.

Coal-Fired Power Plants

The following table summarizes data for each of CILCO's three coal-burning plants:

Station and Unit	Size of Unit (Mw)	In-Service Date	Dependable Capacity When Burning Environmentally Acceptable Fuel (Mw)	
			Actual 1980	Projected 1981
Duck Creek				
Unit 1	390	1976	379	379
E.D. Edwards				
Unit 1	131	1960	75	128
Unit 2	268	1968	193	240
Unit 3	358	1973	341	341
R.S. Wallace				
Unit 3*		1939		
Unit 4*		1941		
Unit 5†		1949		
Unit 6		1952		
Unit 7	322**	1958	132**	135**

*shut down except for emergencies
†this unit is a turbine serviced by 2 boilers
**totals for units 3-7

Eighty-one percent of this capacity, or 948 Mw, has been installed since 1967. A second 400 Mw unit will be added to the Duck Creek Station after 1987.

Emission Standards

CILCO's Wallace and E.D. Edwards stations must meet the Illinois Air Pollution Control Standards that limit sulfur dioxide emissions to 1.8 pounds of $SO_2/10^6$ Btu because they are located near Peoria, Illinois, a major metropolitan area. CILCO's Duck Creek Unit must meet the more stringent sulfur dioxide requirement, 1.2 lb. $SO_2/10^6$ Btu, to comply with the 1971 federal New Source Performance Standards (NSPS).

In 1978 the Wallace Station met air quality emission standards at Units 6 and 7 by reducing their generating rate. Operating permits have been issued for these units with the understanding that they will run at a reduced capacity and burn cleaned high-Btu, low-sulfur, low-ash bituminous coal. Units 1 and 2 have been retired, and Units 3 through 5 were shut down and will be fired only during emergencies.

The E.D. Edwards Station, whose operation will be examined in more detail below, has implemented a plan for meeting emission standards by burning cleaned coal and raw low-sulfur coal in all units and using flue gas conditioning to enhance particulate collection in all boilers.

In order to satisfy the 1.2 lb. $SO_2/10^6$ Btu NSPS, CILCO installed a four-module FGD system which started operating in July 1978 on Duck Creek Unit 1. This system, used in combination with coal cleaning, allows Duck Creek to meet the emission requirement while burning high-sulfur Illinois coal.

Coal Cleaning Policy

The coal purchased by CILCO from Freeman United Coal Mining Company in Illinois and from Northwest Coal Corporation in Colorado is cleaned. Coal mined in Illinois has been cleaned since 1950 to improve its general quality. Coal cleaning can usually reduce the sulfur content of an Illinois coal from between 5 and 6 percent to between 3 and 4 percent. CILCO demanded that Northwest Coal Corporation's coal be cleaned in order to reduce its ash content and make its quality more consistent. Any reduction of sulfur is a side effect, as Northwest's coal is already a "compliance" coal (0.7 percent sulfur). Because coal from Northwest Coal Corporation must travel 1,339 miles from Colorado to Illinois, the transportation-cost savings that result from reduced tonnage are significant.

Although the quality of coal from CILCO's other western contractor, Westmoreland Resources, is creating problems in the utility's boilers, cleaning was not stipulated for this coal. Cleaning it, notes Robert G. Herren, Manager of General Services for CILCO, would probably result in a worse product by adding more moisture to the coal, and perhaps concentrating the already high sodium content, which can cause slagging and fouling in the boilers.

E.D. EDWARDS STATION UNIT 3

Each of the three coal-fired units at the E.D. Edwards Station, which is located just south of Peoria, Illinois, must meet a sulfur dioxide emission standard of 1.8 lb $SO_2/10^6$ Btu. The E.D. Edwards Station began operating in 1960, and the last unit, 3, started up in 1972. Unit 3 can generate 341 Mw net when burning either design fuel or cleaned low-sulfur coal from Northwest Coal Company.

Unit 3 was originally designed to burn Illinois coal. However, in anticipation of having to meet the 1.8 lb $SO_2/10^6$ Btu standard with a coal that would burn well in the boiler, CILCO examined 35 other eastern and western suppliers. Initially CILCO decided to fire Unit 3 with low-sulfur sub-bituminous coal from Westmoreland's Montana reserves. However, the high sodium concentration of this coal produced serious slagging and fouling when it was burned. The coal also created operating problems in the precipitators, and as a result CILCO was unable to satisfy Illinois' particulate requirements. Therefore CILCO switched to using the Montana coal in Units 1 and 2, and used the Illinois coal that these two units had been burning in Unit 3.

Units 1 and 2 were able to meet sulfur dioxide standards while burning the Montana coal. To meet particulate standards, CILCO reduced their maximum operating load.

CILCO used Illinois coal in Unit 3 until January 1, 1979, the date on which it had to comply with the 1.8 lb $SO_2/10^6$ Btu emission standard. It then switched to Northwest Coal Corporation's Colorado coal.

Central Illinois Light signed a 15-year contract with a Northwest Coal subsidiary, Western Slope Carbon Inc. on October 30, 1979, for deliveries beginning in January 1980. This agreement provides for 600,000 tons annually of high-Btu, low-sulfur cleaned coal from Northwest Coal's Hawk's Nest Mine near Somerset, Colorado.

The specifications set forth in CILCO's agreement with Northwest Coal Corporation for coal delivered to Unit 3 of E.D. Edwards are:

Moisture — 7.5% maximum

Ash — 8.0% maximum

Sulfur — 0.7% maximum

Btu/per lb — 12,400

Although CILCO shifted from Illinois coal to cleaned low-sulfur Colorado coal, operating personnel at Unit 3 said that the Colorado coal burned as well as or better than the Illinois coal. It is unusual for any coal—if different in quality from the fuel originally intended—to burn more efficiently than the coal for which the boiler was designed, says Leo Grigsby, Fuel and Contracts Manager. However, continues Grigsby, this unexpected boiler performance is not necessarily due to the fact that Northwest Coal Corporation's coal is cleaned, but probably due to its being of higher quality than Freeman United's Illinois coal.

Benefits and Problems of Burning Cleaned Coal

According to CILCO, burning cleaned coal not only helps the company meet emission standards, but also improves boiler performance and reduces transportation and pulverizer-operation costs. Since the cleaned coal particles from Northwest Coal are 2 inches to 0 in size, as opposed to its raw coal sizes of 5 inches to 0, much less enegy is required of the pulverizer. And since cleaning reduces the amount of extraneous material in the coal by 35 percent, transportation costs decrease.

CILCO noted only one problem related to cleaned coal use: the additional moisture on the coal causes it to freeze in winter. CILCO has tested a freeze-proofing agent on shipments from Northwest Coal. However, Grigsby indicates that the tests to date have not shown freeze-proofing to be of any benefit.

The effectiveness of electrostatic precipitators (ESPs) in particulate removal is affected by the use of any low-sulfur coal, cleaned or raw, because the attraction of particulates to the ESP decreases as a coal's sulfur content decreases. To increase particulate collection at the E.D. Edwards Station, CILCO is injecting sulfur trioxide into the flue gas.

The use of baghouses instead of precipitators can solve this problem, but Grigsby points out that there is not enough space at the E.D. Edwards Station to install them. In addition, flue gas conditioning is less expensive.

Transportation

CILCO's experiences in transporting coal illustrate the complexities and ironies of the present energy and conservation problems in the United States. Northwest Coal's Western Slope Carbon preparation Plant in Colorado is 1,339 miles from the E.D. Edwards power plant in Illinois. A 73-car unit train carrying approximately 7,300 tons of coal makes, on the average, one trip a week to deliver coal to the E.D. Edwards station. To get to Illinois, this train must use the tracks of three different railroads and cross the mountains through the Tennessee Pass in Colorado at an altitude of 11,000 feet with five engines pulling and nine pushing. The amount of fuel consumed by these engines is significant: 50,000 to 60,000 gallons of fuel oil for each round trip. CILCO's contract for Westmoreland coal from Montana also requires long-distance shipping, as the supplier is 1,290 miles away. This compares to a 106-mile trip for coal that CILCO buys from Freeman United's mines in Illinois.

The rate for the delivery by unit train of Northwest Coal Corporation coal to CILCO, established December 5, 1979, is now $22.44 per ton. Such a cost is a factor to be considered by utilities or industries planning to switch from burning oil or gas to burning coal. CILCO is now protesting this rate as excessive.

Coal Costs

CILCO is unable to measure the savings attributable to burning cleaned coal from Northwest Coal Corporation, as opposed to raw coal, because before the use of western coal, the company burned cleaned Illinois coal. Since cleaning Northwest Coal's fuel decreases ash and mining wastes while concentrating Btu's, and also reduces its size, CILCO assumes that pulverizer-operating and transportation costs are reduced.

CILCO's costs for cleaned coal from Northwest Coal Corporation in dollars per ton or per million Btu were unavailable for publication.

In December of 1979, the company paid an average of $36.76 per ton (including transportation costs) for coal from Freeman United, Westmoreland, and Northwest Coal, and the use per million Btu amounted to 178.4¢

NEW ENGLAND ELECTRIC SYSTEM
25 Research Drive
Westborough, MA 01581

Power Plant Profiled:	Brayton Point Station; Units Nos. 1-3 Somersett, Massachusetts

Coal Use:	640,000 tons (12/79 through 7/80)

Total Electrical Capacity (1980):	4,441 Mw
Sources:	Two steam electric generating stations; Four internal combustion stations; One gas-turbine peaking unit Oil-fired units: 24% of capability Coal or oil-fired units: 40% of capability
	One pumped storage facility 14% of capability
	14 hydroelectric stations 13% of capability
	Four jointly-owned nuclear stations 9% of capability
Peak Load (Winter 1980):	3,140 Mw
Reserve Capacity:	29.3% of total electrical capacity
Growth of Electrical Demand:	Prior to 1979: estimated 3.1% Goal for 1996: estimated 1.9%
Service Area:	4,500 square miles in Massachusetts, Rhode Island, and New Hampshire. 194 cities and towns served in tri-state region.
Number of Customers (1980):	1,055,363

Reguatory Commission(s):   Massachusetts Department of Public Utiliti●
                            New Hampshire Public Utilities Commission
                            Rhode Island Public Utilities Commission

Revenues: Total operating revenue (1980):
                            1,090,101,000
                            Fuel Purchases:
                            $531,414,000
                            (49% of total operating revenue)

Rates (1980):   Residential: 7.13¢/Kwh
                            Commercial: NA
                            Industrial: NA
                            Average: NA

Subsidiaries: Massachusetts Electric Company, MA
                            Narragansett Electric Company, RI
                            Granite State Electric Company, NH
                              (retail electric companies)
                            New England Power Company
                              (generates and/or purchases
                              electricity for retail subsidiaries)
                            New England Power Service Company
                              (construction, engineering
                              services, fuel procurement)
                            New England Energy Incorporated
                              (oil/gas exploration, shipping)

New England Electric System (NEES), a Massachusetts utility holding company, has major innovative plans for the use of coal at its fossil fuel power plants in the near future. The company's Brayton Point Generating Station, the largest fossil-fuel plant in New England, is being converted from burning oil to coal, and most of its coal will be cleaned. By 1982 NEES plans to generate 42 percent of its energy using coal, and by 1996, 49 percent.

NEESPLAN

During the late 1960s and early 1970s the company used imported oil to generate most of its electrical power because of the low cost of the fuel at that time. By 1979, 73 percent of the company's power supply was generated by imported oil; the remainder, by domestic oil, nuclear power, and hydroelectric generation. No coal was used at any of the NEES generating stations. As a response to the ever-increasing cost of foreign oil and the company's desire to limit the peak-demand growth of electrical power, "NEESPLAN," a fifteen-year energy plan, was developed. The major goals of NEESPLAN are to reduce use of imported oil to 10 percent of the company's energy needs, in part by burning coal, and to limit peak-demand growth to an average of no greater than 1.8 percent per year in order to reduce the need for new power plants.

NEESPLAN has many strategies in addition to the major conversion of oil-fired power plants to coal. The program includes the development of NEES' domestic oil and gas supplies, the encouragement of customer conservation, the completion of nuclear plants now under construction, and the development of alternative energy sources such as wind, solid-waste and wood combustion, and small-scale hydroelectric power. The company projects that NEESPLAN will save more than 380 million barrels of foreign oil and nearly $1.4 billion in the next 15 years.

By 1982 and 1996, NEES hopes to diversify its energy sources as follows:

	Coal (%)	Oil (%)	Nuclear (%)	Hydro (%)	Alternative Energy Sources (%)
1982:	42	37	14	7	0
1996:	49	10 (foreign) 6 (domestic)	25	6	4

Overall, three power plants may eventually be converted from oil to coal. The company has recently converted Units 1, 2 and 3 of the Brayton Point Generating Station. The company estimates that the conversion at Brayton Point will save approximately 12 million barrels of oil per year by burning about 3 million tons of coal per year. NEES is also considering converting the generating units at the Salem Harbor (Salem, Massachusetts) and South Street (Providence, Rhode Island) stations.

BRAYTON POINT GENERATING STATION

The Brayton Point Generating Station, located in Somerset, Massachusetts, consists of four generating units capable of producing 1,615 Mw of electrical power. Units 1, 2 and 3, which produce 1,150 Mw, can burn either coal or oil. The 465 Mw Unit 4 is designed to burn only oil. The following table describes Units 1, 2 and 3:

Unit	Size of Unit (Mw)	In-Service Date
1	250	1963
2	250	1964
3	650	1968

History of Coal Use

Units 1, 2 and 3 were originally designed to burn coal and did so through 1969. In 1969 the units were converted to burning less expensive foreign oil. In 1974 and 1975, following the Arab oil embargo, Units 1, 2 and 3 burned coal on an emergency basis, and were then converted back to lower-sulfur oil. However, NEES began studying the feasibility of permanently converting Brayton Point to coal. After lengthy negotiations with the U.S. Department of Energy, the U.S. Environmental Protection Agency, and the Massachusetts Department of Environmental Quality Engineering, the company was given permission to burn coal under a "delayed compliance order"* from December 1979 through July 1980. NEES purchased 640,000 tons of cleaned coal on the spot market and burned it in Units 1, 2 and 3 during this eight-month period. Sixteen coal companies supplied the cleaned coal at an average spot-market price of $33 per ton. NEES reports that 3 million barrels of oil and $24 million in fuel costs were saved by the use of this coal.

NEES switched back to burning oil at Brayton Point while the construction for permanent conversion was completed. Unit 1 began burning coal again in early April of 1981; Unit 2, at the end of June 1981; Unit 3, in mid-November 1981. Together they burn about 245,000 tons a month (about 3 million tons a year).

---

*A delayed compliance order is a regulation issued by EPA that allows a utility generating station to temporarily exceed emission limits while additional pollution-control equipment is being installed. In the case of Brayton Point, additional particulate-control equipment was installed as part of the conversion agreement.

Pollution Control

The Brayton Point station is subject to sulfur dioxide emission regulations set by the State of Massachusetts. Its coal-fired boilers cannot burn a fuel that exceeds 1.21 pounds of sulfur per million Btu on a 30-day average, or 2.31 pounds of sulfur per million Btu during a 24-hour period. To comply with this regulation, NEES will burn mostly cleaned coal with a sulfur content of less than about 1.5 percent.

Sources of Coal and Coal Specifications

Damon Lawrence, Manager of Coal Supply at NEES, visited 16 coal companies during the summer of 1980 to study the quality of their fuel and the efficiency of loading and shipping operations. After analyzing his findings, the company signed contracts with five companies from Virginia, West Virginia, and Pennsylvania.* All of the coal burned at Brayton Point must meet the following specifications:

Sulfur—1.21 lb/10^6 Btu maximum;* 0.80% minimum

Ash—10% maximum

Moisture—8% maximum

Btu per pound—13,000 minimum

*Equivalent to 1.57% sulfur at 13,000 Btu per pound

NEES did not specify that this coal be pre-cleaned. However, Lawrence noted that most deep-mining companies supplying coal to the utility own preparation plants and use them to remove mining wastes, improving the raw coal in order to meet buyers' ash and sulfur specifications.

The use of low-sulfur western coal has been considered by NEES. However, the Btu content of this coal is too low to assure efficient burning in the Brayton Point boilers, which were designed for a higher-Btu fuel. In addition, transporting coal from the western United States to

*The companies are A.T. Massey Coal Company, United Coal Company, Island Creek Coal Company, Johnstown Coal & Coke Company, and Benjamin Coal Company.

Massachusetts presents enormous contracting and logistical difficulties, including the necessity of using several different railroads.

NEES is also considering some form of investment in a coal mine as part of its future energy strategy. Further company investigations will determine whether the benefits and risks justify the utility's owning coal reserves.

Benefits and Problems of Burning Cleaned Coal

The major advantage of coal cleaning, states Lawrence, is that it enables more coal reserves to meet the ash and sulfur requirements set by NEES. The sulfur dioxide emission standard for older boilers in Massachusetts limits the pounds of sulfur per million Btu in the fuel, rather than the pounds of sulfur dioxide per million Btu in the boiler's flue gases. NEES meets this requirement by burning mostly cleaned coal with a uniform sulfur content.

According to Lawrence, coal cleaning may or may not reduce the sulfur content of the coal. He points out that when large amounts of ash and only small amounts of sulfur are removed by physical coal cleaning, the percentage of sulfur per pound in the cleaned fuel may actually increase. However, he believes that coal cleaning can produce a more uniform, consistent fuel that will not vary as widely in sulfur content as raw coal. Thus the emission standard can be achieved with greater ease when cleaned coal is burned.

The major problems with cleaned coal, according to Lawrence, are the high percentage of fine particles in the fuel and the frequently high moisture content of washed coal. Lawrence believes that the high "fines" content of cleaned coal may generate more dust than raw coal does. The company studied this problem during the summer of 1981. Dust control is achieved primarily by water sprays during transportation, unloading, and storage. Lawrence notes also that some of the Btu value of the raw coal is lost in the preparation plant wastes.

NEES has not quantified the costs of burning cleaned coal as compared with raw coal. However, some savings were accomplished in pulverizer operations at Brayton Point. Because 50 percent of the cleaned coal the station buys is less than one inch in size, it creates less wear on the equipment used to reduce the size of coal before burning.

Transportation

Approximately two shiploads of coal per week will be unloaded at the Brayton Point station. The company projects that in 1981 approximately 1 million tons of coal will be delivered; in 1982 and the years beyond, approximately 3 million tons of coal will be delivered each year.

All of the coal purchased by NEES from its five suppliers will be shipped a total of 800 to 900 rail and sea miles from mine to generating station. The coal will be hauled by rail from mines in central Pennsylvania, West Virginia, and southern Virginia to loading ports on the Chesapeake Bay. There the coal will be transferred to ocean vessels that will deliver it to Brayton Point in Massachusetts.

Three different railroads will transport coal for NEES. Lawrence states that NEES decided to use several railroads not only to generate competition for favorable shipping rates, but also to increase shipping reliability. The company is building a coal-fired ship that will be capable of hauling 2.2 million tons of coal per year. NEES will also ship a portion of its coal purchases in vessels it leases.

Coal Costs

NEES signed long-term and short-term contracts with its five coal suppliers. The price of the coal it purchases ranges from $30 to $40 per ton.

PUBLIC SERVICE COMPANY of INDIANA, INC.
1000 East Main Street
Plainfield, IN 46168

Power Plant Profiled:	Gibson Station; Units Nos. 1-4 East Mt. Carmel, Indiana
Coal Use:	12,407,760 tons (1979)-coal comes from mines mainly located in Indiana and Illinois
Total Electrical Capacity (1980):	4,951 MW

Sources:	six coal-fired stations - 98.3% of power production
	one hydroelectric station- 1.6% of power production
	15 rapid-start internal combustion generating units (oil) - 0.1% of power production
Peak Load (Summer 1980):	3,896 MW
Reserve Capacity (1980):	21.3%
Growth of Electrical Demand:	3.79% - 1979-80 (4% per year projected)
Service Area:	North-central, central, and southern Indiana. Services 69 of 92 counties in Indiana - estimated population: 1.9 million
Number of Customers (1980):	533,144
Regulatory Commission:	Public Service Commission of Indiana
1980 Total Operating Revenues:	$645,700,000
Fuel Purchases:	$276,012,000 (42% of total operating revenue)
Rates: 1980	Residential 4.32¢/Kwh Commercial: NA Industrial: NA Average: 3.39¢/Kwh

The Public Service Company of Indiana, Inc. (PSI), is a typical midwestern utility that uses high-sulfur and high-ash Illinois Basin coal to generate almost all of its electrical power (98.3 percent in 1980). Although PSI does not specify that its coal be cleaned, the company reports that in 1980 between 66 and 100 percent of the fuel burned at each of its coal-fired power plants was cleaned, in order to meet sulfur, ash and Btu specifications.

The 2,600 Mw Gibson Generating Station profiled in this study is PSI's largest power plant. About 66 percent of its coal was cleaned in 1980, and three preparation plants, including the Monterey Coal Company No. 2 Preparation Plant (see Monterey Coal Company profile), supplied fuel to the station.

Coal-Fired Power Plants

The following table summarizes the data on PSI's generating stations:

Station	Unit	Size of Unit (Mw)	In-Service Date	Fuel Pre-cleaned(%)*
Gibson	1	650	1975	65.8
	2	650	1976	
	3	650	1978	
	4	650	1979	
	(5)		(1982)†	
Cayuga	1,2	500 each	1970-72	86.3
Gallager	1-4	160 each	1958-61	100.0
Wabash	1	365	1968	95.0
River	2	125	1956	
	3-6	100 each	1953-54	
Edwardsport	1,2	45 each	1944	100.0
	3	75	1951	
Noblesville	1,2	45 each	1950	90.0

*As of August 1980
†Unit 5 is expected to become operational in 1982

PSI generally uses long-term contracts for coal that comes from mines located mainly in Indiana and Illinois. In 1979 PSI purchased its coal through contracts of the following types: long-term, 88.6%; spot-market, 11.4%; captive production, 0.0%. Almost all (97.1 percent) of PSI's cleaned-coal purchases are made by long-term contract, with the remainder made on the spot market. The company receives cleaned coal from a total of 11 preparation plants.

Policy on Coal Cleaning

According to Gene Aimone, PSI's Fuel Procurement Supervisor, the company has historically set coal specifications for sulfur, ash, moisture, and Btu's per pound that allow for the use of either cleaned or raw coal, as long as the fuel can meet the company's requirements for a given generating station. However, PSI is committed at present to the use of Illinois Basin coals, which usually require cleaning in order to reduce their inherently high sulfur and ash content.

PSI elected not to burn low-sulfur western coal because of the high cost of transporting it to generating stations in Indiana. It is Aimone's opinion that coal cleaning will become more necessary for utilities such as PSI because of the decreasing quality of present and future coal reserves.

GIBSON GENERATING STATION

The Gibson Generating Station, located in East Mt. Carmel, Indiana, is a four-unit, 2,600 Mw coal-fired facility. Each of the generating units is rated at 650 Mw. Another 650 Mw generating unit, now under construction, is to be completed by the fall of 1982.

The units operating at the Gibson station are required to meet the Indiana sulfur dioxide emission standard of 5.8 pounds of SO_2 per million Btu. This emission standard is less stringent than most of the standards set for utilities profiled by INFORM. PSI complies with the standard by burning cleaned or raw coal with a maximum sulfur content of 4 percent.

PSI is constructing a flue gas desulfurization system for the fifth Gibson generating unit. This scrubber will enable the company to burn high-sulfur Illinois Basin coal and meet the federal New Source Performance Standard for sulfur dioxide emissions at the new unit. Aimone reports that PSI will also use cleaned coal (in combination with scrubbing) at Unit 5, but the specifications for the coal to be burned have not yet been determined.

PSI has not developed any cost analyses of combining coal cleaning and scrubbing. However, the company believes that it will save money by burning cleaned coal in conjunction with scrubbing, because the scrubber will not have to remove as much sulfur dioxide from the flue gases as it would if raw coal with a higher sulfur content were burned.

Sources of Coal and Coal Specifications

Sixty-six percent of the approximately 5,820,000 tons of coal burned at the Gibson station was pre-cleaned in 1980. Three preparation plants provided fuel for the station. Of Gibson's coal needs, 87.5 percent are met by long-term contracts with coal suppliers, and 12.5 percent by purchases on the spot market.

All coal suppliers must meet the following specifications assigned to the Gibson generating units:

Sulfur—4% maximum

Ash—15% maximum

Moisture—16% maximum

Btu per pound—10,000 minimum

The coal need not be cleaned as long as it meets these specifications. The company reports that all boilers can operate at their rated capacities when using either cleaned or raw coal. PSI does not blend raw and/or cleaned coals burned at the Gibson station, although coal suppliers sometimes blend coals sold to PSI in order to meet the fuel specifications.

Benefits and Problems of Burning Cleaned Coal

Aimone sees "the elimination of excess foreign material" (mining wastes and coal ash) as the major advantage of using cleaned coal. He states that PSI could not burn much of the Illinois Basin coal supplied to the utility if most of the clay, shale and other non-combustible materials were not removed. He says that coal cleaning improves the Btu content and results in more efficient boiler performance. Sulfur removal is useful but less important to Aimone, because the Gibson power plant only hs to meet relatively relaxed sulfur dioxide emission standards.

The major disadvantage of cleaned coal, according to Aimone, is the problem of fine coal particles produced by the process. These particles present transportation, storage and fuel-handling difficulties for the utility.

PSI claims to have no "significant" problems with dust control during the transportation of coal shipments. During unloading, storage and handling, the company uses vacuum-type equipment to control dust emissions and reports no difficulties controlling dust during "normal" operations.

Transportation

In 1979, 85 percent of all coal deliveries to the Gibson station were by railroad (4.8 million tons), and 15 percent were by truck (0.85 million tons). Normally the station receives two 7,500-ton trainloads of coal a day, six days a week, and truck deliveries of varying amounts five days a week. The Monterey Coal Company No. 2 Preparation Plant (one of three preparation plants supplying Gibson with cleaned coal) uses unit trains to ship the coal to the generating station.

Coal Costs

PSI's costs for cleaned coal at the Gibson station (as delivered) were between $22.50 and $29.50 per ton during the first six months of 1980 (between $1.06 and $1.34 per million Btu).

PSI claims to be unable to quantify the costs of burning cleaned versus raw coal. The only saving noted by PSI that resulted from burning cleaned coal with a lower ash content was a decrease in the costs of handling and disposing of boiler ash.

Advanced Physical Coal Cleaning Processes

AMES LABORATORY
Iowa State University
Department of Chemical Engineering
Ames, IA 50011

Oil agglomeration is a coal cleaning technology that can successfully improve the quality of extremely fine-size coal that is difficult to clean by other processes. The cost of removing water from the coal cleaned using this technology is lower, as it does not require expensive thermal drying of the final product.

Background

Ames Laboratory is a facility devoted to basic energy research, located on the campus of Iowa State University (ISU). Of the laboratory's funds, 98 percent come from the U.S. Department of Energy (DOE) and 2 percent from the State of Iowa.

In 1974 ISU was allocated $3 million by the Iowa State Legislature to research the problems associated with the mining and use of Iowa coals. Since these coals contain high percentages of sulfur (up to 8 percent), ISU decided to study various methods of coal sulfur reduction and removal. After examining a number of physical and chemical coal cleaning technologies, the university selected three promising methods of coal cleaning for further development: 1) froth flotation, a conventional coal cleaning method; 2) oxydesulfurization, a method of chemical coal

271

cleaning; and 3) oil agglomeration, an advanced physical coal cleaning technology.

Research of the various coal cleaning technologies, including oil agglomeration, began in 1974 at the ISU Energy and Mineral Resources Research Institute. Dr. Thomas D. Wheelock of the Chemical Engineering Department carried out this research, with funding provided by the State of Iowa for the first three years. In 1977 sponsorship shifted to the federal Energy Research and Development Administration (ERDA), and the project was transferred to the Ames Laboratory at ISU. Wheelock has continued as director of the project, and it has been funded primarily by DOE since 1977. ISU has spent approximately $250,000 on the R&D of oil agglomeration since 1974. Thus far, no patents have been issued for the oil-agglomeration processes developed at ISU.

The oil-agglomeration project at ISU has emphasized laboratory research designed to improve the basic process. Most of the work has been conducted at the lab site, using small equipment. Variables such as coal type, oil type, oil concentration, particle size, particle concentration, and slurry pH have been studied.

Since 1974 Ames Laboratory has tested coal from many different mines in Iowa, Illinois, Kentucky and Pennsylvania. In one experiment using an Iowa coal ground to -400 mesh (a very small size), Ames Laboratory reported an 88 percent reduction of the raw coal's inorganic (pyritic) sulfur after cleaning by oil agglomeration. The cleaned coal contained only 0.6 percent inorganic sulfur, the ash content was reduced by 63 percent, and about 93 percent of the burnable organic matter was recovered. Other coals from Iowa and western Kentucky tested by Ames Laboratory showed reductions of inorganic sulfur of between 70 and 88 percent, a reduction of 60 to 74 percent in the ash content of the raw coal, and a high recovery (96 to 99 percent) of the burnable matter.

Process Description

This advanced physical coal cleaning technology, originally developed during World Wars I and II, uses the difference in surface properties between the organic and inorganic components of raw coal to separate the coal from waste elements. This principle is the same as that used in the conventional froth-flotation process. Inorganic waste components, with the notable exception of pyritic sulfur, tend to attract water (are hydrophilic), while the valuable organic components usually repel water (are hydrophobic).

In the oil-agglomeration process, fuel oil is added to a batch of water-soaked, extremely fine-sized (-400 mesh) coal particles. The batch, or suspension, is agitated so that the oil disperses into small droplets which selectively coat the organic, water-repelling particles. These oil-coated particles then stick together, forming larger clumps (called flocs or agglomerates) of clean coal. The flocs are separated from the unagglomerated, water-attracting mineral waste particles—which are suspended in water—by screening and/or centrifuging the suspension. The agglomerates, the final product, include both the coal and the added oil droplets—which increase the Btu value of the cleaned fuel.

Pyritic sulfur, unlike other inorganic wastes, is not naturally water-attracting. Thus a special process must be used to remove this impurity. Wheelock reports that the removal of pyritic sulfur from coal particles depends in part on the pH of the original coal-and-water slurry. When the pH of the slurry is between 7 and 11, pyritic sulfur can be successfully removed from the raw coal. This more alkaline slurry may alter the surface characteristics of the pyritic sulfur and cause this material to become water-attracting instead of water-repelling. In order to create the alkaline conditions that favor pyritic sulfur removal, raw coal is chemically pretreated, usually with sodium carbonate.

Several other important variables that affect oil agglomeration have been discovered by Ames Laboratory. The amount of coal used in relation to oil and the intensity of agitation can affect the size of agglomerates and the ease of water removal. When relatively small amounts of oil (less than 5 percent of the coal weight) are used in conjunction with fast agitation, small agglomerates are formed that must be centrifuged to remove water. However, when larger amounts of oil (more than 5 percent of the coal weight) are used with less agitation, larger agglomerates are formed that can be dewatered easily on screens. Also, the size of raw coal particles cleaned by the oil-agglomeration process influences the effectiveness of the technology. Larger, -60 mesh particles cannot be as completely cleaned as smaller, -400 mesh particles. Finally, the type of oil used in the procedure can help or hinder the cleaning process. Heavier oils are more effective than light oils in reducing the pyritic sulfur and ash contents of Iowa coals.

Benefits

According to Wheelock, oil agglomeration has three advantages over the conventional coal cleaning technologies. First, it is one of the few coal cleaning technologies that can be used to physically clean and recover extremely fine-size coal. Froth flotation, states Wheelock, does not generally separate fine-size clean coal and waste materials as well. The fact that oil agglomeration works effectively on very small coal particles is an important advantage, because some kinds of coal contain mineral matter that can only be removed by grinding the coal to a very fine size to unlock the individual mineral particles.

Another advantage of oil agglomeration is that the energy requirements of preparation plants could be reduced by using this process. Since the large agglomerates can be dewatered by screens and the smaller agglomerates by centrifuging, the energy-intensive process of thermal drying can be avoided, potentially reducing both capital and operating and maintenance costs. Finally, this technology may present fewer waste problems. Liquid wastes should be minimal because most process water will be reclaimed and recycled.

Problems

The principal disadvantage of oil agglomeration is the relatively high cost of oil. Even though common fuel oils can be used and the oil can be burned along with the cleaned coal, the cost per Btu is higher for the oil than for the coal (as of July 1980 the average U.S. prices of light fuel oil and coal in cents per million Btu were 562.8 and 137.5, respectively). Typically, the amount of oil required is 8 to 10 percent of the weight of the dry coal.

The need for chemical pretreatment to separate coal and pyrite particles is another disadvantage. Without chemical pretreatment to alter the slurry pH, water-repelling pyritic sulfur particles are agglomerated almost as readily as the coal particles. Wheelock reports that most of the recent research effort at Ames Laboratory has been directed toward developing better methods of pretreatment.

Wheelock maintains that it is difficult to discuss the transportation advantages and disadvantages of coal cleaned by oil agglomeration. He says that problems may arise, depending on the method of agglomeration and the physical characteristics of the agglomerates—smaller agglomerates present more handling difficulties than larger ones. In some cases, agglomerated coal may be mixed with larger coal sizes and handled in a conventional manner. In other situations, free-flowing agglomerates of uniform size can be produced, and such agglomerates can be transported and handled as if they were fluids.

Future Development and Marketing

Wheelock notes that research on oil agglomeration at Ames Laboratory is now directed toward improving coal Btu recovery, increasing the separation of sulfur and ash, and reducing the amount of oil needed in the process. Although ISU does not plan to commercialize the process itself, it will continue with its basic R&D, subject to the availability of funding from DOE or other sources. Wheelock reports that in the long term, technical assistance will be provided to others who wish to develop and use the process commercially.

Wheelock states that oil agglomeration seems particularly well suited for commercial applications where a coal-and-oil mixture can be burned. It appears that such a mixture can be burned in some utility and industrial boilers that were originally designed to burn fuel oil; large-scale firing tests are under way to demonstrate this concept. Also, oil agglomeration may be an economical way of recovering discarded fine-size coal from preparation plant wastewater and from old slurry ponds. Thus Wheelock sees the process as a useful adjunct to other methods of coal cleaning.

Oil agglomeration has been investigated in a number of countries, including Australia, Canada, India and West Germany. At least two commercial plants have been built and operated in West Germany, although just one is now operating, and only intermittently. Wheelock reports that the high cost of oil has discouraged commercial operations.

Wheelock envisions a possible future role for oil agglomeration in combination with physical coal cleaning technologies that cost less (e.g., float-sink separation) and with flue gas desulfurization (FGD). He believes that it would be advantageous to combine coal cleaning with partial scrubbing of boiler flue gases to avoid the thermally inefficient and costly practice of reheating the gases. This practice is usually necessary when all flue gases are passed through an FGD system.

Projected Costs

According to Wheelock, the cost of oil agglomeration will be higher than the cost of conventional coal cleaning. This increased cost will greatly depend on the price of the fuel oil used. Although Ames Laboratory has not made a detailed cost estimate of oil agglomeration, it refers to an economic analysis written by Capes, Smith and Puddington

(1974) for theoretical 100 and 200 ton per hour oil-agglomeration plants. The results of the Capes study indicate that oil agglomeration is sensitive to both coal and oil costs, and relatively insensitive to variations in plant capacity, capital costs and fixed costs. For a theoretical 200 ton per hour plant, Capes estimates that total process and fixed costs would be $1.50 per ton of raw coal processed (1974 dollars).

HELIX TECHNOLOGY CORPORATION
266 Second Avenue
Waltham, MA 02254

Clean Pellet Fuel is a new coal-and-limestone fuel that emits low levels of sulfur dioxide when it is burned. The pellet also burns with less smoke than raw coal, and is a stronger material, being less susceptible to breaking apart during transportation and storage. However, Clean Pellet Fuel is not an advanced physical coal cleaning technology per se. It can best be described as a "hybrid" technology that contains features of coal cleaning, flue gas desulfurization (FGD) and coal gasification. For purposes of organization, it is being included in this section on advanced physical technologies.

Background

Helix Corporation, through its McDowell-Wellman Division, has developed a coal-and-limestone pellet known as "Clean Pellet Fuel" (CPF) at the company's Cleveland Research Center. (In 1978 Helix acquired the McDowell-Wellman Company, which had initiated coal pelletization research in 1975.) In May of 1980, Dravo Corporation bought portions of Helix which had been developing the pellet, and Helix extended its license on the pellet to Dravo. However, Dravo abandoned its research facilities in 1981, and all of the processing rights for the pellet fuel have been retained by Helix.

Coal pelletization is a descendant of the technologies which were developed in the 1950s to make pellets out of powdered iron ore[*] and out

---

[*] Powdered iron ore (recovered from low-grade taconite) is commonly pelletized in order to allow for easier shipping and smelting. McDowell-Wellman has developed a process to use an aluminum waste product, "red mud", which contains anywhere from 30 to 60 percent iron, to form a pelletized iron product. This technical process was a forerunner to coal pelletization.

of power plant fly ash. McDowell-Wellman, which had been involved in iron ore pelletization and oil shale research, began serious R&D of coal pelletization in 1975. Although the chemistry of the coal and iron ore pelletization processess is different, the machinery used to manufacture the pellets is basically the same. Thomas Ban, McDowell-Wellman's Vice President of Research and Development, developed CPF in 1975, and it is protected by U.S. patent No. 4,111,755.

Initially McDowell-Wellman became involved with coal-and-limestone pelletization because: 1) coal seemed to be a fuel that would be burned in increasing quantities; 2) profits from iron ore pelletization were on a downward trend; and 3) a research engineer employed at McDowell-Wellman had previously conducted small-scale graduate work with CPF.

R&D costs for Helix's development of Clean Pellet Fuel are estimated by the company at between $1 and $2 million. The Ohio Department of Engergy has funded about one half of the R&D costs, with the remainder provided by internal corporate sources.

Helix has conducted commercial combustion testing of CPF at the Massillon State Hospital and the National Lime and Stone Company in Ohio. More extensive stack-gas analyses and combustion testing of CPF have since been performed at the Ohio State Reformatory in Mansfield, Ohio.

The testing at the Ohio sites has supplied promising data on the sulfur-reduction capabilities of CPF. A 2.12 percent sulfur pellet emitted an average of 0.91 pounds of sulfur per million Btu (0.91 lb S /10^6 Btu) after it had been burned directly (0.34 lb S /10^6 Btu in best test case), and a 2.03 percent sulfur pellet that had been gasified* produced an average of 0.42 lb S /10^6 Btu (0.12 lb S /10^6 Btu in best test case). In contrast, a 2.00 percent sulfur raw coal (13,000 Btu) would emit about 1.50 pounds of sulfur per million Btu. Helix also reports that smoke and other particulate levels were significantly lower during the combustion tests as compared with "normal coal firing operations."

Process Description

This technology is somewhat different from the other advanced

---

*Gasification tests were performed with a "Wellman-Galusha" gas-producing device, where CPF was heated to produce a gas which was then burned in a secondary furnace.

physical coal processes described in this report, as it involves features of traditional coal cleaning technologies, FGD, and coal-gasification processes. The purpose of coal-and-limestone pelletization is to cause the sulfur component of the raw coal to react with the lime component of the fuel and remain in the ash waste product while the pellet is being burned or gasified. In this manner, sulfur products such as sulfur dioxide are not released as gaseous effluents, but are transformed into disposable solid calcium sulfate in the ash.

Four steps are involved in the production of CPF. First, predetermined amounts of coal, limestone and several carbonate additives are measured. Then the materials are ground to -28 mesh and mixed. Next, a moistened blend of coal and limestone is balled and compacted to form "green pellets." Finally, the green pellets are either "carbonized" or "pyrolyzed" in order to attach the sulfur from the coal component to the limestone.

In carbonization, the green pellets are heated to 1,600°F to 2,000°F in a traveling grate furnace. The carbonized pellet is later burned as a fuel. During the carbonization process, part of the green pellet's volatile matter is burned, creating by-product heat that can be used to generate steam, possibly for cogeneration. About 80 percent of the green pellet's sulfur content remains in the ash when the carbonized pellet is finally burned.

In pyrolysis, the green pellets are burned in a sealed, oxygen-free traveling grate furnace, where gases emitted from the green pellets are partially burned at 900°F to 2,000°F. Low-Btu gases and liquids are generated as by-products and can be recovered as usable fuels. Over 90 percent of the sulfur content of the green pellet is retained in the ash residue when pyrolysis is employed.

Benefits

Christopher Brody, a research engineer at the McDowell-Wellman Cleveland Research Center, says that the positive environmental attributes of CPF are the greatest advantage of the fuel when compared with raw coal. He adds that, unlike other physical coal cleaning technologies, CPF enables sulfur dioxide emissions to be controlled without the need for FGD systems. Pilot testing, states Brody, suggests that the federal New Source Performance Standard (NSPS) for sulfur dioxide can be met by burning CPF pellets that have been gasified.* If CPF is burned directly, only a small portion of the flue gases would require scrubbing in order to meet the 1.2 pounds of SO_2 per million Btu NSPS.

Volatile material removed during the production of CPF eliminates smoke and tar problems during combustion, maintains Brody. The swelling and sticking properties of some raw coals are also eliminated by using CPF, which allows for "satisfactory combustion and gas production." Finally, Helix claims that CPF is both stronger and more resistant to degradation by weathering than raw coal, facilitating easy transportation, handling and storage.

Problems

According to Brody, there are several problems with CPF. Since limestone is mixed with coal to form CPF, there is approximately 50 percent more non-burnable ash in the pellet than in raw coal. This ash takes the form of bottom-ash in boilers when the fuel is burned, and the disposal of this ash may create solid-waste difficulties. In addition, limestone lowers the Btu value of the fuel, since it is not burnable. Also, an additional 10 to 15 percent of the fuel's Btu value is used up during carbonization, which necessitates using the heat produced to generate steam—to be used or sold for cogeneration—in order to make the fuel economically competitive with other fuels. However, Brody contends that the problems associated with increased ash content are offset by the savings in sulfur dioxide control equipment that is not needed when CPF is burned.

Brody also reported fuel-ignition problems during combustion testing. These problems were related to the low percentage of volatile material present in CPF, and the company is alleviating these difficulties by modifying the furnaces using the pellets at the test sites.

A final difficulty with CPF is the 25 percent moisture content of the fuel, which it acquires after carbonization, when the pellet is water-quenched in order to prevent oxidation and further loss of heating value. Brody notes that Helix is investigating alternative quenching technologies in order to control this problem.

---

*However, Brody did not comment on CPF's ability to meet the more stringent *revised* NSPS.

Future Development and Marketing

As of early 1981, Helix completed pilot-plant testing of CPF, although "some refinements" remain necessary for both boiler engineering and fuel chemistry. Brody expects the next development step to be the design of a CPF demonstration plant. This project will be initiated if, after the company completes its review of CPF's commercial potential, it decides to continue its R&D of the fuel. Brody believes that Helix will build CPF plants for potential customers, and leave the day-to-day operations to the plant owners.

Brody envisions several potential markets for CPF. He feels that CPF is well suited to industrial applications that can use the gases generated from the fuel. Because the amount of gas that can be produced at present from CPF is limited by the small size of gasification equipment that is on the market, utilities will not make use of the fuel. However, the development of larger "circular grate furnaces," which would permit utilities to generate up to 400 Mw of electricity, could make CPF a viable fuel alternative. CPF can also be burned in either industrial stoker furnaces or chain-grate stoker furnaces that are commonly used in hospitals, schools and prisons.

A final market option for CPF, according to Brody, is in the field of fuel conversion. He points out that there are technical difficulties encountered when oil- or gas-fired boilers are converted to coal. If CPF is gasified, the fuel can be substituted for oil or gas without the need to build another boiler facility. Although he does not offer any figures to support this claim, Brody believes that conversion to burning gas generated from CPF is less expensive than conversion to burning coal.

Brody sees CPF as having a potential role in the future U.S. energy scene, although the fuel "is not going to take over the market." He feels that the escalating price of fuel oil will enable alternative fuels such as CPF to become cost-effective. Brody concludes that CPF is a viable coal-based fuel, and at this time "it is as good as other ideas."

Projected Costs

Helix did not provide the estimated capital and annual operating and maintenance costs for a CPF plant. The company did offer the following estimates of raw coal and CPF costs in dollars per million Btu:

raw coal, $1.20 per million Btu; clean pellet fuel, $2.00 per million Btu (for carbonized CPF; costs are estimated at less for pyrolyzed CPF).

Brody maintains that burning a low-Btu gas produced from CPF is less costly than building and operating an FGD system, although he did not supply cost-comparison data.

OAK RIDGE NATIONAL LABORATORY
Oak Ridge, TN 37830

The dry magnetic coal cleaning technologies may offer effective and relatively inexpensive alternatives to the conventional wet cleaning of fine coal particles. In a dry process, the fine coal is cleaned without the use of a coal-and-water slurry, and costly and energy-intensive water-removal equipment is therefore unnecessary.

Background

The Engineering Technology Division of Oak Ridge National Laboratory has been conducting research and development which focuses on two magnetic coal cleaning technologies known as "High Gradient Magnetic Separation" (HGMS) and "Open Gradient Magnetic Separation" (OGMS), under the direction of E. C. Hise.

Oak Ridge had its beginnings during the World War II Manhattan Project, when its principal mission was the production and chemical separation of the first gram quantities of plutonium. Today the goal of this national energy laboratory is to develop technologies for the efficient production and use of energy. Fossil fuels have been studied at Oak Ridge for approximately six years. The facility is operated by Union Carbide Corporation's Nuclear Division for the U.S. Department of Energy (DOE).

Oak Ridge's magnetic-separation research has been funded primarily by DOE's Division of Fossil Fuel Extraction. The Tennessee Valley Authority (TVA) has also supplied funding for the High Gradient Magnetic Separation tests. The coal for the tests has been contributed by TVA and Pennsylvania Power & Light Company.

Oak Ridge initially began R&D of the magnetic coal cleaning technologies in order to "demonstrate the feasibility and develop to commercial use a process for the magnetic separation of the inorganic contaminants of pyritic sulfur and ash-forming minerals from *dry* crushed coal." In 1976 Oak Ridge's dry magnetic coal-separation program was initiated. Thus far, research into High Gradient Magnetic Separation has been conducted by Oak Ridge at both Oak Ridge itself and Sala Magnetics, Inc., located in Cambridge, Massachusetts, and owned by Allis Chalmers. E. C. Hise reports that a U.S. patent application has been filed for the Open Gradient Magnetic Separation equipment developed at Oak Ridge.

In 1978 Oak Ridge tested separate batches of 100 mesh to 200 mesh size coal (Kentucky No. 9 seam) with HGMS equipment. These tests indicated that HGMS can remove about 60 percent of the pyritic sulfur from the raw coal, and return 90 to 95 percent of the Btu value. When this data was compared with conventional gravity coal cleaning achievements calculated by Oak Ridge, HGMS was found able to remove equivalent proportions of pyritic sulfur, although the process was not as effective as gravity-separation in removing ash (only 50 percent ash-removal was accomplished in one series of tests). INFORM's sample of coal cleaning plants suggests that conventional coal cleaning can remove about 60 percent of a raw coal's ash content. However, similar testing of another coal (Pennsylvania Lower Freeport) with HGMS equipment suggested that both sulfur and ash reduction could approach those attained by the gravity-separation processes.

Process Description

Both the High Gradient Magnetic Separation and the Open Gradient Magnetic Separation technologies are based on the differences between the magnetic susceptibility of coal and waste particles. HGMS involves capturing pyritic sulfur and other inorganic coal impurities that are weakly attracted to a magnetic field in a magnetic separator. Non-magnetic organic coal particles pass through the separator as cleaned coal. Pyritic sulfur and other minerals that are usually present in raw coal in the form of fine, widely diffuse particles are, in general, partially liberated from raw coal by crushing and/or breaking before the use of HGMS.

HGMS has been used successfully with other substances. Since 1969 fine, weakly magnetic impurities have been removed by the HGMS process from Kaolin Clay, a material often used in paper manufacturing. Subsequently, magnetic-separation technology has been applied not only to the field of coal cleaning, but also to wastewater treatment, iron ore beneficiation, and particle-removal from gas streams.

The HGMS system employed by Oak Ridge is composed of a matrix of stainless steel mesh or filaments surrounded by a solenoid (an electromagnet) that creates a magnetic field around each filament or element of the mesh. The slightly magnetic pyrites and inorganic particles are drawn into the field and are trapped on the mesh or on the filaments, while the clean coal passes through the system. The pyrites and inorganic materials are later released from the mesh or filaments by simply turning off the magnet. In order to clean coal by the HGMS process, coal must be in the form of fine particles. The particles can be suspended in air or slurried in water as they pass through the mesh; however, Oak Ridge's work involves the use of the dry technique.

In late 1979 Oak Ridge began testing HGMS equipment, at the Sala plant, that was capable of continuously separating wastes from clean coal at the rate of one ton per hour. Continuous separation is necessary for the full-scale commercialization of this process, as batch processing only moves a small amount of coal through the equipment for later evaluation. According to Hise, preliminary tests indicate that continuous HGMS can "compare favorably with batch tests being run concurrently."

Open Gradient Magnetic Separation is a similar coal cleaning process that is now at a less-developed stage than HGMS. It utilizes the magnetic forces that attract pyritic sulfur and inorganic materials, and the magnetic forces that repel organic coal particles away from a magnetic field. In this process, a stream of crushed coal is fed into a cavity under the influence of a specially designed magnetic field that deflects clean coal particles and attracts refuse particles, producing a "spectrum" of material. This spectrum can be split in order to separate the stream into product coal and refuse. OGMS is a dry process in which the raw crushed coal is fed through the magnetic field either by vertical free fall or on a vibrating tray.

At the present time, coal is cleaned by OGMS with a laboratory machine capable of processing small amounts of coal, and a superconducting magnet capable of processing only one third of a ton per hour. However, Hise points our that OGMS can potentially perform a more precise separation between coal and coal wastes than can HGMS, since OGMS enables the particles to be split into discrete streams of clean coal and wastes. Furthermore, HGMS can only capture particles that are weakly magnetic. Oak Ridge's research confirms this hypothesis, finding that ash reduction is typically more successful with OGMS. However, more research is required before its suitability for industrial use can be assessed.

Benefits

Hise sees both HGMS and OGMS as being potentially effective methods for cleaning fine coal. In the past, fine coal particles have been discarded by preparation plants, but today's economics make this practice prohibitively expensive.

Since HGMS and OGMS are both dry coal cleaning processes, several problems related to moisture are avoided. (Excess moisture in coal can add to shipping costs, cause coal to freeze during storage, and reduce the heating value of the fuel.) In addition, centrifuging and expensive, energy-intensive thermal drying are minimized.

The research at Oak Ridge is based on the premise that dry coal cleaning processes will ultimately be more economical than wet processes. Preliminary cost estimates (to be discussed) suggest that a 200 ton per hour HGMS plant can clean coal at lower operating costs than a comparable "conventional" plant using thermal dryers.

Problems

Difficulties with HGMS and OGMS include (1) the less-than-average ash-removal capability suggested by some of the previously cited research of HGMS by Oak Ridge; (2) the tendency of coal particles smaller than 10 microns to agglomerate during the HMGS procedure, interfering with the separation of clean coal from wastes (if these extremely small particles are not removed prior to cleaning); and (3) material handling as a result of "technical" difficulties.

Future Development and Marketing

Oak Ridge is developing an engineering and economic evaluation of HGMS, in conjunction with DOE and TVA. Flow sheets, equipment lists, and cost estimates of both HGMS and conventional wet coal cleaning processes are being completed for use in a 200 ton per hour fine-coal cleaning plant. The final phase of this project, to be completed by the end of fiscal 1981, will be the development of an actual engineering design and cost estimate for an HGMS pilot plant. According to Hise, if the continuous separation testing and engineering and economic evaluations are "favorable," and once technical material-handling problems are solved, Oak Ridge, DOE, and Sala Magnetics will construct and operate a 10 to 20 ton per hour HGMS pilot plant at TVA site (most likely at Paradise, Kentucky) with the utility's participation (see TVA profile).

Future R&D of OGMS at Oak Ridge will be directed toward building magnets capable of processing coal "at rates of many tons per hour." Hise concludes, "Since this process offers the potential of being simpler, cheaper, and more selective than the HGMS, we are now experimentally corroborating the calculations and will attempt to develop it as a commercial process."

Several other companies and research facilities have expressed an interest in magnetic coal cleaning technologies. Stackpole Corporation, General Dynamics, Babcock and Wilcox, and Los Alamos Scientific Laboratory have made recent inquiries concerning applications of HGMS and/or OGMS. A Brazilian coal company has also requested assistance in developing magnetic coal cleaning for a high-ash Brazilian coal that has been difficult to clean using conventional coal cleaning methods.

Hise sees a twofold advantage in using cleaned coal as a utility fuel; it has environmental advantages and it may be a cheaper fuel to burn than raw coal. He concludes that HGMS and/or OGMS will play an important role in future coal cleaning applications, assuming that economic benefits can be firmly established.

Projected Costs

Oak Ridge provided preliminary cost estimates for a 200 ton per hour HGMS facility that processes high-sulfur, high-ash coal (Western Kentucky No. 9 seam) as follows: operating costs of $3.78 per ton of clean coal, with a capital investment of $23 million, as compared to operating costs of $4.97 per ton of clean coal and a capital investment of $16 million for an average conventional plant with thermal drying. The high capital costs of HGMS are primarily due to the investments for the magnets and air-classification equipment. It may well be possible to make substantial reductions in the cost of this equipment as the process matures.

OTISCA INDUSTRIES, LTD.
P.O. Box 127
Salina Station
Syracuse, NY 13208

The Otisca Process uses a non-toxic organic liquid to clean coarse and fine coal without the need for the water-and-magnetite suspensions employed by conventional plants. In this manner, high percentages of inorganic sulfur and ash are removed from the raw coal without adding surface moisture, and therefore water-removal equipment is not necessary with this process. Otisca's developers claim that the process offers lower capital and operating costs than the conventional coal cleaning technologies.

Background

Otisca Industries, Ltd. (originally Chemical Comminution International), was founded in August 1972 by Dr. D. V. Keller, Jr., a chemist, and C. D. Smith, a mechanical engineer, with the goal of developing new technologies for the preparation of coal. The company's interest in coal stemmed from past research conducted by Dr. Keller at Syracuse University, where he held the position of Professor of Material Sciences. The "Otisca Process" was invented by Keller and Smith in 1972 and is protected by a number of U.S. patents, principally No. 4, 173,530.

In 1972, after the Otisca Process had been successfully demonstrated in laboratory tests, approximately $250,000 was raised through a private-venture capital stock issue to support efforts toward its commercialization. According to Keller and Smith, the results of the laboratory tests "projected a potential for commercial efficiency and economics in the areas of fossil-fuel extraction and upgrading."

Smith reports that 63 percent of the total sulfur, 70 percent of the inorganic (pyritic) sulfur, and 87 percent of the ash was removed from a raw Pennsylvania coal during laboratory testing of the Otisca Process in 1973. The Btu values increased from 9,128 per pound to 14,009 per pound, and the surface moisture of the raw coal was reduced from 6.50 percent to 2.07 percent during the cleaning process.

The following table provides more recent data, furnished by Otisca Industries from a larger demonstration plant that cleans coal using the Otisca Process. The characteristics and waste by-products are listed for raw coal, coal cleaned by a conventional water-and-magnetite process, and coal cleaned by the Otisca Process. Otisca Industries did not specify the type of FGD system or water-and-magnetite coal cleaning system used in formulating the comparative data presented in this table.

Effectiveness of the Cleaning Plant

Coal Characteristics	Raw Coal	Cleaned Coal	Decrease/ Increase
Sulfur (%):	6.2	5.8	4.2
Ash (%):	24.8	18.0	11.5
Btu per pound:	10,700	11,500	12,700
Waste By-Products			
Pounds of FGD wastes generated per year*	14,260	11,310	8,060
Pounds of R-11 emissions per year +	0	0	0.1 to 0.5

*Figures for sludge from FGD scrubbers assume the generation of 10 pounds of sludge per 1 pound of sulfur in the raw coal.
+R-11 is the organic liquid used as a cleaning medium in the Otisca Process.

The Otisca Process represents the major undertaking of the company at the present time.* To date, Otisca Industries has spent approximately $9 million on research and development of this technology. The funds for R&D have come from major collaborative efforts with American Electric Power (AEP) and Island Creek Coal Company, as well as from "many small testing grants supplied by industry." According to Otisca Industries, no government funding has been used for the R&D of this process.

Promising results of laboratory and batch testing conducted by Otisca Industries from 1972 to 1973 led to Island Creek Coal Company's financing a 20 ton per hour pilot plant at the company's North Branch Mine in Bayard, West Virginia. In January 1977 the pilot-plant equipment was moved to a larger site in Florence, Pennsyl-

---

*Other technologies that have been or are now being studied include chemical comminution, tar-sand hydrocarbon-solvent extraction, and two other advanced coal cleaning process (Otisca "T" and "B" processes).

vania, where Otisca is continuing operations and equipment development.

In 1977 Otisca Industries also signed a contract with AEP to design, construct and operate a 125 ton per hour demonstration plant at AEP's Muskingum Mine located near Beverly, Ohio. This $6.7 million facility is a small coal preparation plant that can process coal from the mine to provide clean coal for testing and possibly for commercial use. The 125 ton per hour production goal has been achieved, although some minor equipment modification has been necessary. (The company did not specify what this equipment modification was.)

Process Description

The Otisca heavy-liquid coal-beneficiation process is a variation of the conventional water-and-magnetite coal cleaning technologies. In this process a heavy organic liquid, trichlorofluoromethane (CCl_3F), otherwise known as Refrigerant-11 (R-11), is used as a cleaning medium instead of the conventional suspension of water and magnetite. Pyritic sulfur, ash and other wastes are removed by gravity separation in a static bath of R-11, producing a cleaned coal product. The surface moisture present on water-repelling coal particles is transferred to water-attracting waste particles by adding certain compounds (not specified by the company) to the R-11 before gravity separation. The cleaned coal and the refuse are then heated to approximately 100°F, which evaporates almost all of the R-11 present on the surface of the particles. The R-11 vapors are then condensed and recycled. Smith reports that R-11 has a low toxicity and is chemically stable, non-explosive, non-flammable and odorless.

Benefits

According to Otisca Industries, its process can clean both coarse and fine raw coal efficiently without the need for costly size-classification equipment. In fact, coal particles as small as 200 mesh, which is too small for most conventional coal cleaning systems to clean, can be cleaned by the Otisca Process. Also, the absence of water in the cleaning circuits, combined with the use of compounds to transfer surface moisture from the coal to the wastes, produces a clean coal product with virtually no surface moisture. Therefore thermal dryers, as well as the other equipment used to remove water from cleaned coal in conventional preparation plants (e.g., vacuum filters, dewatering screens and water clarification equipment), are not necessary.

Otisca Industries also claims that its process can dramatically reduce corrosion problems in plant equipment, as the chemical properties of R-11 help avoid the acidic (low pH) and corrosive environments found in water-based coal cleaning systems.

Otisca Industries foresees no problems in transporting coals cleaned by the Otisca Process.

Other advantages of the Otisca Process over the conventional coal cleaning methods, according to Keller and Smith, include the following: 1) more complete pyritic sulfur and ash removal, 2) higher Btu yield, 3) relatively dry refuse for disposal, 4) a sealed system,* and 5) favorable economics. This final advantage can be attributed to the need for less plant equipment (water-related circuits) and lower maintenance costs as a result of reduced corrosion.

Problems

The only disadvantage that Otisca Industries sees in its process, in comparison to the other pre-combustion coal cleaning technologies, is that Otisca's is a new technology and therefore it has high initial costs and requires some technical "debugging." Otisca Industries claims that there are no environmental hazards related to R-11 emissions and feels that the limited emissions of R-11 (see table earlier in this profile) from an Otisca plant can be considered a positive trade-off against the increased amount of waste sludge from flue gas desulfurization (FGD) that would be generated if raw or conventionally cleaned coal is treated with FGD.

Future Development and Marketing

Otisca Industries plans to commercialize the Otisca Process by building and operating plants, and by selling the rights to build and

---

*Since the Otisca plants use recycled R-11, there are no water emissions from the plant, such as the "blackwater" streams that are common in open preparation plants.

operate such plants to other contractors. Keller and Smith see economic advantages in combining pre-combustion coal cleaning and FGD, although they note that such advantages are site-specific. The company sees the Otisca Process as being most applicable for use in the utility steam-coal market, although it is considering other possible markets, such as coke manufacturing by the steel industry and the gasification and liquefaction technologies that may require cleaned coal. The foreign markets for coal cleaned by the Otisca Process are viewed as being potentially profitable for the company, and international patents have been obtained. (The company did not mention particular countries.)

Projected Costs

Otisca Industries has provided cost data (in 1979 dollars) on the Otisca Process for a projected 400 ton per hour plant with a ½ inch to 0 feed of raw coal.

Capital costs (does not include coal-handling equipment): $6,000,000 or $15,000 per ton of raw coal per hour

Direct operating costs (labor, fuel, power, materials and maintenance): $1.45 per ton of raw coal

Total costs (at 15% return on investment): $3.65 per ton of raw coal

---

OTHER TECHNOLOGIES

Two other advanced physical coal cleaning technologies may prove to be viable methods of improving the quality of raw coal. However, both processes are further from commercial use than the four advanced physical technologies profiled above, and more research is necessary to determine their usefulness. Electrostatic separation is a dry cleaning process that can remove ash and pyritic sulfur from fine coal particles. Like other dry processes, this technology can save money, as it does not require water-removal equipment. The Kintyre Process is an Australian invention. Its developers claim that it can remove very significant amounts of ash and sulfur. However, little is known about the specifics of the process, and a recent article in an Australian journal has questioned the credibility of Kintyre's developers.

Electrostatic Separation

Bechtel Corporation has reported limited success with several electrostatic coal cleaning technologies. Generally these technologies clean raw coal by separating individual particles of coal, pyritic sulfur, and ash-forming substances, by using their differing levels of electrical conductivity. No commercial electrostatic separators have ever been built in the United States, although two pilot plants were built in Germany before and during World War II.

Promising results of experiments with an electrostatic technology have been reported by Bechtel. The technology, known as "triboelectric separation," can decrease the pyritic sulfur and ash found in raw coal. Bechtel reported the following pyritic sulfur and ash content for the raw and cleaned coal:

Pyritic sulfur—in raw coal, 2.08%; in cleaned coal, 0.81%.
Ash—in raw coal, 7.95%; in cleaned coal, 5.39%.

Another advantage of triboelectric separation over the other physical cleaning technologies is that the process can separate ultrafine coal particles (particles as small as 100 mesh) from wastes. Also, since triboelectric separation is a dry coal cleaning technology, the clean coal does not need to be thermally dried, leading to capital and operating and maintenance savings in a commercial operation. According to Bechtel Corporation, triboelectric separation may have potential as a commercially viable coal cleaning technology if sufficient R & D of the process continues.

The major problem with triboelectric separation is equipment wear. Bechtel Corporation concludes that wear-resistant materials need to be developed to replace stainless steel parts, which have proven to be prone to corrosion from coal and waste particles.

The principle of triboelectric separation is simple. A "triboelectric" charge is generated by friction when coal particles are allowed to slide down a metal chute or pass through an air classifier. One study reported that coal particles can acquire a net positive charge and pyrite and ash-forming shale particles a net negative charge when passed through a triboelectric separator. The positively charged coal particles can then be split from the negatively charged wastes.

Drum-type electrostatic separation is an older but less successfully developed technology. Drum separation differs technically from triboelectric separation in that raw coal particles are exposed to an electrical field on a rotary drum, leading to the separation of clean coal from the wastes. The most conductive pyrite particles lose their charge first, and are ejected by gravity and centrifugal forces. The coal particles are poor conductors and stick to the drum, and at a later point they are removed with a scraper. Particles of intermediate conductivity are removed at a point on the drum between the clean coal particles and the wastes. This point can be preset on the drum by the use of splitting devices that direct the clean coal and waste streams.

Bechtel offers several possible explanations for the overall lack of success in developing drum separators. This equipment, because of design constraints, can only effectively clean a limited amount of coal per hour (approximately 40 tons per hour) unless several drums are used as modules. Also, drum separators are very sensitive to the differing properties of raw coal, and separation quality is difficult to maintain. A final difficulty with this process is that dust explosions are a hazard associated with drum separators, creating dangerous operating conditions for laboratory and plant personnel.

Advanced Energy Dynamics, Inc. (AED), of Natick, Massachusetts, has recently developed a coal cleaning system based on drum-type electrostatic separation, which may show more promise than past efforts. AED claims that its system is 32 to 77 percent cheaper than conventional coal cleaning technologies, and almost twice as effective in removing sulfur from raw coal. Finely pulverized coal has been fed into iron ore drum-separator modules,* modified for coal, which have removed 37 to 68 percent of the sulfur and 51 to 59 percent of the ash from a variety of raw coals during tests conducted by AED. The company has solved problems associated with drum separators and dust explosions by physically trapping the fine coal particles between the drum and a "boundary" layer of air. AED also reports that its system can be used to clean coal either at mine sites (like conventional coal cleaning technologies) or at the site of coal use (e.g., a utility power plant).

---

*Each module can process 3 tons of coal per hour, and AED reports that a battery of modules can clean over 1,000 tons of coal per hour in a full-size preparation plant.

Kintyre Process

Kintyre Enterprises (also known as Energy Recycling Corporation), an Australian Company, has announced the development of an advanced physical coal cleaning process and a burner system designed to be used with the coal cleaned by this process. The company claims that "most" of the sulfur and ash found in raw coal can be eliminated by this process. The Kintyre Process has been developed from a German World War II missile-launching technology. Raw coal is fed into the meeting point of two high-velocity gas streams and "explodes," due to the immediate expansion of the water contained within the coal, as the coal "churns" in the gas streams. Ash, pyritic sulfur, and moisture are removed at this stage, although the company has not described how the removal is accomplished. Organic sulfur is removed at a "later stage," which has also not been explained by Kintyre. The company claims that Australian coals cleaned by the Kintyre Process contain 15,000 to 16,000 Btu per pound, with less than 0.1 percent ash, 0.4 percent sulfur, and 0.5 percent moisture. However, the characteristics of the raw coal used by Kintyre in its experiments were not provided by the company, which makes an analysis of the effectiveness of the process difficult.

An article in a recent Australian journal (*ERT*, September 1980) has raised several questions about the Kintyre Process. According to the author, Alan J. Lloyd, no "evidence or product samples" have been available to back up the claims made by Kintyre. Although a Texas utility, the Lower Colorado River Authority, has shown an interest in the Kintyre Process, Lloyd doubts that the process will receive serious attention unless more is known about the fuel and how it is produced.

Research and Development Organizations

ELECTRIC POWER RESEARCH INSTITUTE
3412 Hillview Avenue
P.O. Box 10412
Palo Alto, CA 94303

The Electric Power Research Institute (EPRI), research arm of the electric power industry, was chartered in 1972 to develop a broad technological program for improving the production, transmission, distribution and use of electric power in an environmentally acceptable manner.* With increased utility reliance on coal, EPRI has directed its attention to finding the cleanest and cheapest ways to use this fuel.

By 1985, our nation's utilities will spend $1 billion per month on coal versus $750 million per month in 1980. At the same time, the quality of raw coal will have steadily deteriorated with the introduction of new mining techniques that bring up less accessible coal, which contains greater amounts of impurities. EPRI believes that as the utilities become more dependent on coal, and as the quality of the raw coal deteriorates, research and development to improve coal quality by coal preparation will become more important. The objective of its research and development program is to improve the technology in order to reap the benefits that coal cleaning offers in power production, and to make the utility industry aware of all the possible benefits.

---

*EPRI is funded entirely by contributions from utilties. To be a full member, a utility must pay an amount of money which is prorated on the basis of its power production.

Dr. Sheldon Erlich, former manager of EPRI's Fluidized Combustion and Coal Cleaning program, states that "utilities and utility customers are the ultimate beneficiaries of coal cleaning, and that is why the industry must hone this art to a science." According to EPRI, research and development in coal cleaning could result in a more uniform, lower-ash fuel that would decrease the frequency of conventional pulverized-coal boiler breakdowns due to slagging and fouling. By eliminating ash, coal cleaning also maximizes the Btu content of the coal, reducing the weight of the coal for a given number of Btus, thereby presenting potential savings in shipping costs. EPRI also views coal preparation as a method of decreasing the cost and complexity of meeting the environmental regulations placed on coal burning power plants.

However, Randhir Sehgal, Manager of Coal Cleaning at EPRI, stresses that in most cases the importance of coal cleaning for utilities is as an ash rather than as a sulfur-reduction technology. He warns that coal cleaning is not a cure-all for pollutants, stating that "coal cleaning does not make pollutants disappear, it separates them from the coal." However, capturing sulfur in a solid form rather than dispersing it into the air as sulfur dioxide allows the utility to contain the pollutant in a controlled waste disposal area.

According to Sehgal, the benefits of coal cleaning that improve boiler performance are difficult to quantify. As long as coal companies can sell their coal without cleaning, he says, they will not clean it. Kenneth Clifford, former Manager of Coal Cleaning at EPRI, states that the utility industry "wants to sit down with numbers and be able to look at the extra costs paid for cleaned coal versus the savings in boiler operations and maintenance costs." Without the quantitative data on the costs and benefits of coal cleaning, a utility's decision to stipulate cleaned coal in its boilers is often based on a vague notion about the advantages of cleaned coal. However, recent tests, some of which are EPRI sponsored, are seeking to quantify these costs and benefits, and Sehgal believes that the number of utilities demanding cleaned coal will therefore increase.

Program and Technology

EPRI's coal cleaning research and development is under the aegis of the Coal Combustion Systems Division. At present EPRI has about 20 coal cleaning projects under way. Most are conducted by outside consultants under the auspices of two full-time EPRI staff members. In 1980 and 1981, EPRI's coal cleaning budget was between $4 and $5 million. The 1982 budget is expected to be approximately the same. Since government funded research and development on coal cleaning has recently been cut back, EPRI's program is expected to become increasingly important.

EPRI's coal cleaning research has four principal concerns: 1) to improve existing coal cleaning technology, 2) to determine the effects of cleaned coal on the cost of power production, 3) to develop methods to get an even lower ash fuel, and 4) to design instrumentation and automation for preparation plants. More specifically, research objectives are aimed at:

1. Developing improved crushing systems to produce coals with a controlled size distribution:

2. Exploiting gravity differences and surface chemistry more effectively;

3. Reducing the costs of dewatering and drying;

4. Developing process control strategies and automated unit operations;

5. Determining the impact of various contaminants in coal on the costs of power production;

6. Developing techniques such as oil agglomeration and froth flotation to produce a coal with a very low ash content.

EPRI has built a $10 million coal cleaning test facility that can process up to 20 tons of raw coal per hour. This facility, located in Homer City, Pennsylvania, began operating in August of 1981. EPRI uses it to test research results on a scale larger than lab or bench scale. The facility, according to EPRI representatives, will also help the utility industry improve the economics of coal cleaning by developing superior flow sheets for "steam" coal, testing flow sheet design before installation, and testing various unit operations. According to Sehgal, however, the emphasis of the facility will not be on testing individual plant flow sheets but rather demonstrating to the utilities that coal can be cleaned effectively and economically.

The design of the facility is flexible enough that almost any plant flow sheet in use today can be set up by reconnecting the various units in the plant (heavy-media cyclones, water-only cyclones, tables, flotation cells, classifying cyclones, centrifuges, and thickeners). The test facility was designed for easy sampling and continuous monitoring of the various process parameters. EPRI believes that after four or five years of testing, a large coal cleaning data base will be available to assist the industry in designing preparation plants on a more scientific basis. EPRI would also like to acquire a pilot combustor for its research and development and use it to make comparisons between the combustion characteristics of raw coal and those of coal cleaned to the various quality levels. EPRI believes this is important in improving the utilities' understanding of the benefits of using cleaned coal.

EPRI's most exciting prospects in coal cleaning involve the improvement of dewatering equipment, the more effective recovery of magnetite (for reuse) used in heavy-media equipment to clean fine coal, and the development of an on-line testing apparatus for various coal constituents. EPRI has several ongoing projects in each of these areas.

The objective of one of the dewatering projects being carried out at the University of California in Berkeley, is to reduce the moisture content of 28 mesh to 0 fines after conventional dewatering, from the current level of 22 percent to about 10 percent. The successful development of this technology will reduce water consumption in coal preparation plants and reduce the dependence on thermal drying, thus encouraging coal cleaning. According to Sehgal, removing ash but then adding water to coal does not improve the fuel and thermal drying is too expensive a method of removing the moisture added by washing.

A project to facilitate the recovery of magnetite in fine cleaned coal and fine refuse is being conducted by Eriez Magnetics for EPRI. On the average, 25 pounds of magnetite are lost for each ton of fine coal processed. To replace this magnetite would cost $1.00 to $1.25 at the current price of 4¢ to 5¢ per pound. Eriez Magnetics is attempting to reduce the amount of magnetite lost per ton of fine coal processed to 5 pounds, saving 20 pounds per ton, and thus generating significant savings. The make-up magnetite required per ton would then only cost 20¢ to 25¢.

EPRI also has several projects aimed at developing an on-line testing apparatus that can determine certain characteristics of coal (its sulfur, ash, moisture and Btu content, and/or the percentage of all its other constituents) which are important to both preparation plant and power plant operators. Preparation plant operators must deal with coal whose composition may differ by mine, seam, or even truckload. On-line testing apparatus would enable the operators to adjust the specific gravity of heavy-media flotation units thereby insuring that the precise coal specifications set by its customers will be met.

One new device for coal analysis that EPRI is perfecting is the Continuous On-line Nuclear Analysis of Coal (CONAC). The most important advantage of CONAC over the traditional American Society of Testing Materials (ASTM) laboratory analysis is speed. According to EPRI the results are as accurate as the ASTM procedures for determining the constituents of a coal. By immediately knowing the composition of the coal being fed to a boiler, the operator can compensate for fuel variations and their effects on the efficiency, maintenance requirements and power costs of the boiler.

Outlook on Research and Development

Refinements are still needed in conventional coal cleaning operations. The historical approach to coal cleaning has been empirical. It has been viewed as an art rather than a science, especially for the coals used in electric power production. According to Sehgal, the techniques and machinery used in coal cleaning have not been significantly improved within the last 15 years—with the possible exception of the new freon-based Otisca cleaning process (see Otisca Industries profile).

Clifford believes that coal cleaning is currently less sophisticated than flue gas desulfurization (FGD). Although coal cleaning has existed longer, it is not a technology whose use has been mandated by the government. Consequently, its refinement has been haphazard and slow. On the other hand, FGD was mandated ten years ago for most new boilers, and the best chemical engineers and financial resources were devoted to fine tuning the technology. The coal cleaning industry owes much of what technological sophistication it has to the metallurgical industry, which cleans its coal extensively for use in the coking process.

UNITED STATES DEPARTMENT OF ENERGY
Washington, DC 20585

The U.S. Department of Energy (DOE), whose mandate is to increase the development of U.S. energy resources and to encourage the conservation of existing U.S. energy supplies, has placed growing emphasis on coal use in utility and industrial boilers as a means of reducing their consumption of oil. DOE's allocations for research on cleaner and cheaper direct use of coal and on synthetic fuel conversion, has grown over the past few years. In the past year, emphasis has shifted from state-of-the-art improvements to research on longer-range technologies with a higher economic and technical risk—such as cool water slurries and chemical coal cleaning. The Reagan administration has indicated that it intends to radically reduce or abolish DOE. If the department is eliminated, it is unclear which programs will continue under other agencies and which programs will be cut.

DOE maintains that coal quality will be of utmost importance in determining transportation, capital, operating and environmental control costs in coal-fired or retrofitted utility and industrial boilers. The aim of DOE's Fossil Energy Coal Evaluation and Refining program is to develop cost-effective technologies to upgrade the quality of U.S. coals to levels approaching that of fuel oil. The agency views coal cleaning as one potentially important means of achieving this goal.

According to DOE, the demand for high quality coals over the next decade can be expected from: 1) the conversion of oil-fired boilers to coal (approximately 95,000 Mw); 2) the construction of new or replacement coal-fired boilers; 3) industrial users; 4) the development of new fuel forms such as coal-oil and coal-water slurries; 5) the growth of the fledgling synfuel industry; 6) the expansion of slurry transport systems; and 7) the growth of exports. This increase in demand, the agency states, will severely strain our reserves of indigenously "clean" or easily washable coals. Either poorer quality coals will have to be upgraded through improved coal cleaning technologies or coal users will increasingly face uncertain environmental and operating costs.

DOE asserts that two major coal cleaning developments are needed: 1) cost-effective technologies to remove finely dispersed pyritic sulfur and ash from the vast reserves of coal that cannot be cleaned economically by current practices; and 2) methods to remove organic sulfur from coals whose quality is limited largely by its presence. In both cases, finer grinding is needed than is normally practiced commercially. As a result, new and improved methods are required to clean and condition fine coals. After liberating the sulfur, ash, moisture and other impurities from "hard-to-clean" coals, the cleaned fines must be put back into a marketable solid or slurry form by new cost-effective technologies. The agency notes that the variability of coal and the need to adjust processing designs to deal effectively with differences in the characteristics of coal make development of methods to clean and "package" fines "challenging;" the ability to control the quality of the product is critical to the use of advanced coal cleaning technology.

DOE's Coal Preparation Program budget for the fiscal year 1980* was $12.6 million; for 1981 it was $7.2 million. Because of the new administration's stance on government funded R&D, an even lower budget is expected in 1982. The Office of Management and Budget's appropriation pending before Congress is approximately $3 million.

Most of the R&D work in coal preparation sponsored by the federal government is done by the Coal Preparation Division of DOE's Pittsburgh Mining Technology Center (PMTC), located in Pittsburgh, Pennsylvania. As of the Fall 1981, DOE's key focus in its R&D was the analysis of coal cleaning costs and of any savings that would result to users of cleaned coal, giving utilities more reliable data as a basis for their decisions on converting from oil to coal. According to DOE, such studies quantifying the savings associated with burning cleaned coal— as opposed to raw coal—are few.

Program and Technology

DOE's Program centers on advanced technologies to extract as much ash and pyritic sulfur from intermediate and high-sulfur U.S. coals as consistent with reasonable Btu recovery and to develop new unconventional chemical methods to extract 85 to 95 percent of the total sulfur and ash from these coals, thereby obviating the need for flue gas scrubbers on newer utility boilers. DOE is devoting most attention to coals that may be used by currently oil-fired boilers (approximately 95,000 Mw), coals to be used in projected new coal-fired boilers, and coals that now emit the highest amount of sulfur dioxide (i.e., intermediate and high-sulfur midwestern coals). DOE states that it is also directing similar efforts to those lignitic and subbituminous reserves

*The U.S. government's fiscal year runs from October 1 to September 30.

that are expected to be large suppliers for the utility industry.

Sixty-seven percent of DOE's budget for coal preparation goes to contractors doing research for PMTC and 33 percent is used for in-house research and administration. The industrial contractors working on these projects include TRW, United Coal Company, and General Electric. Other tasks are being pursued by the University of Pittsburgh, the Jet Propulsion Laboratory, M.I.T., Grand Forks Energy Technology Center, General Electric and Ames National Laboratory.

The work includes coal sampling, characterization, washability, testing and correlation. According to DOE, this is needed for the development of predictive analytical procedures to project the effect of specific variables on coal cleaning requirements. The effect of washability conditions on trace element distribution is also being studied.

Technology and advanced research includes fine coal cleaning by froth flotation, oil agglomeration, magnetic and special gravimetric processes, and fine coal grinding, dewatering, and slurry formation. Chemical processes that could remove 90 percent of the ash and sulfur from coals are being developed in TRW's fused salt and GE's microwave concepts. These processes have removed 90 percent of the sulfur and ash from Illinois coal in preliminary laboratory tests. (For more information on these and other chemical coal cleaning concepts, turn to Coal Cleaning Technologies, Chemical Process Chapter.)

Process feasibility evaluation, the third major task area, is to confirm the feasibility of removing 90 percent of a coal's ash and total sulfur by processing "tonnage lots" in prototype equipment (smaller than commercial size). Basic facilities for this research exist at Ames Laboratory in Ames, Iowa. The coal and the waste materials will be characterized after processing, in part to determine where the trace elements end up.

The last major task, dissemination of coal cleaning data and evaluation, DOE reports, will test the cleaned coal samples from the prototype equipment under user conditions in the lab and bench-scale tests (ash fouling tests for boilers and synfuel operations), and give these results to interested coal users, coal producers and government officials.

To highlight the findings of its research, DOE plans to choose coal samples from utilities in selected geographical regions, that are currently using (on long-term contract) intermediate and high-sulfur coals, and to perform its own combustion tests on these coals and their variants. DOE expects that the tests will bring attention to any potential advantages of the advanced cleaning technologies. This applies to recurring problems in operating plants as well as to major design and construction planning decisions relating to new plants, new fuel forms, exports or transportation systems.

Outlook on Research and Development

Because oil is more expensive than coal, the total cost of an oil-fired plant is in most cases significantly greater than the total cost of a coal-fired plant, despite the fact that the latter generally has higher capital, operation and maintenance costs over the course of its life. A DOE spokesman states that it is critical for utilities to be aware of this large gap in costs as part of the savings accrued in operating a coal-fired plant could be used to buy cleaned coal instead of raw coal, "making the coal a more uniform fuel approaching oil." In addition, the official states, the extra investment in cleaned coal would often be recouped later by savings in boiler operation and maintenance, transportation and scrubber costs and, in some cases, increases in boiler availability and rated capacity.

Coal preparation technologies are becoming increasingly sophisticated, according to DOE. Until recently, little attention has been paid to coal cleaning as a means of preparing the ultimate feedstock for utility boilers, industrial boilers or synfuel processes. The Agency sees a similar degree of sophistication entering the preparation of coal for utility boilers, where in certin cases, some of the constituents removed during the washing of coal would be reintroduced to balance the acid/base ratio in order to improve the quality of the fuel and avoid slagging and fouling. When the subbituminous coal delivered to Arizona Public Service Four Corners Station in New Mexico was washed in an attempt to upgrade the quality of the fuel, for example, slagging and fouling in these boilers increased.

UNITED STATES ENVIRONMENTAL PROTECTION AGENCY
Washington, DC 20460

Programs and Goals

The Environmental Protection Agency (EPA) is one of the two federal agencies most involved in coal cleaning research and development. EPA's program differs from that of the Department of Energy (DOE) in that it focuses on coal cleaning as a means to achieve environmental goals rather than on the development of new coal cleaning technologies.

Between 1970 and 1979, before the inception of DOE's coal cleaning program, EPA sponsored most of the federal coal cleaning research. From 1974 through 1978, EPA's budget averaged $3 million annually. In 1979, EPA received a special appropriation of about $6 million from Congress for its coal cleaning work, bringing its total budget that year to $9 million. In both 1979 and 1980, EPA received about $750,000, in addition to its $1 million to $2 million Congressional appropriation from DOE, which had taken over most of the funds for coal cleaning research. With the new administration, EPA's funds for coal cleaning research have disappeared; the agency's 1981 budget is less than $250,000 and no funds are budgeted for 1982.

Monies allocated to EPA before 1981 have supported nine coal cleaning research and development projects. Four of these projects studied pollutants that may be categorized as "priority pollutants" under the provisions of the Federal Water Pollution Control Act, as "hazardous wastes" under the Resource Conservation and Recovery Act, or as "hazardous air pollutants" under the 1977 Clear Air Act Amendments or the Toxic Substances Control Act. EPA is seeking ways to minimize the impact of these pollutants on the environment. (For a description of the various pollutants from coal preparation operations, see Coal Preparation Plant Waste Disposal in the appendix.)

EPA's Physical Coal Cleaning Demonstration Program, which is studying Pennsylvania Electric Company's Homer City preparation plant, encompasses four more of the agency's coal cleaning research and development projects. The goals of this three-year program, in which EPA is working together with the Electric Power Research Institute (EPRI) and the Pennsylvania Electric Company (PENELEC) are: 1) to evaluate the feasibility of physical coal cleaning in achieving compliance with sulfur dioxide emission regulations, 2) to gather enough technical data on plant and equipment performance to facilitate the design of other plants, and 3) to compare the economics of physical coal cleaning with that of other sulfur dioxide control technologies. (See PENELEC profile for more information on the Homer City preparation plant and power plant.)

EPA's ninth coal cleaning project, which in 1980 and 1981 was managed by EPA and funded by DOE, is the investigation of the desulfurization of coal through the application of microwave energy. (For more information on this and other chemical coal cleaning methods, see Chemical Coal Cleaning Section.)

These projects fall under the auspices of EPA's Industrial Energy Research Laboratory in Research Triangle Park (IERL–RTP). All nine of these projects are conducted by consultants for EPA, including Battelle Columbus Laboratory, General Electric, the Illinois State Geological Survey, Los Alamos Scientific Laboratory, PENELEC, Tennessee Valley Authority, U.S. Geological Survey and Versar, Inc.

After October 1, 1981 (which marks the beginning of fiscal year 1982 for the government), three EPA coal cleaning projects are scheduled to continue: Los Alamos Scientific Laboratory's trace element characterization and evaluation of coal and coal waste, the Homer City Physical Coal Cleaning Demonstration Program, and TVA cost studies comparing the economies of physical coal cleaning to that of other pollution control technologies.

Future Research and Development

An official with EPA cited two areas of coal preparation that would benefit from further research. A coal cleaning test facility in each of the country's three major coal regions (Appalachia, the midwest, and the west) would help develop the most efficient coal cleaning technologies for the coals produced in each of these regions. (EPRI has just completed a coal cleaning test facility in Homer City, Pennsylvania. For more information on this facility see EPRI profile.) The long term potential of leaching of hazardous trace elements from aging waste disposal piles also deserves study to determine the long term impact of present preparation plant waste disposal methods.

Appendix

Preparation Plant Wastes

The National Coal Association reports that 359 million tons, or about 52 percent of the 691 million tons of coal mined in the United States in 1977, were cleaned to some degree. From this 359 million tons of coal fed to preparation plants, 254 million tons of cleaned coal (36.8 percent of the tons mined) and 105 million tons of waste (29.2 percent of the raw coal feed) were produced.[1] In addition, the congressional Office of Technology Assessment reports that more than 3 billion tons of coal wastes already lie in the 3,000 to 5,000 refuse banks in eastern coal fields alone.[2] Preparation plant operators must therefore contend with a significant waste-disposal problem. Some of the wastes can be toxic and, unless they are disposed of properly, dangerous to the environment. However, several technologies exist to handle these wastes in ways that can minimize environmental problems.

The following chapter gives a general description of preparation plant wastes, their environmental impact, disposal methods and costs, and the government regulations controlling waste treatment. Most of the information was obtained from previously published sources. This chapter supplements the information gathered from INFORM's sample of preparation plants and their waste products. (For data drawn from the sample, see Part II: Findings.)

The Environmental Impact of Preparation Plant Wastes

Five major environmental risks are associated with the disposal of preparation plant wastes.

1) *Acid drainage* at waste disposal sites is the most significant.[3] Acid

leachates, or liquids formed from the wastes, are directly toxic to plant and animal life, and these leachates may contaminate drinking water sources.[4] According to the U.S. Environmental Protection Agency (EPA), about 10,000 miles of streams and rivers in the United States have been damaged by acid drainage from underground and surface coal mines and from preparation plant waste piles.[5] The Appalachian Regional Commission estimates that "combined mining operations" (mines at which both surface and underground mining takes place), mining waste piles, and preparation plants produce about 17 percent of all acid drainage.[6]

The presence of oxygen, water, and sulfur compounds (such as pyrites) in preparation plant wastes results in chemical reactions which can create acidic liquid wastes.[7] In most cases, the higher the sulfur content of the mined coal, the greater the acidity of the liquid wastes.[8]

2) Acid leachates can also dissolve minute quantities of *trace elements* (including heavy metals) such as iron, calcium, manganese, magnesium, mercury and copper.[9,10] These and other trace elements are more readily soluble in water under acidic conditions, and may create toxic hazards when present in sufficient quantities.[11] In addition, some of them can enter the food chain when they are absorbed by lower forms of plant and animal life.[12] Another hazard is posed by trace metals such as arsenic, cadmium, nickel and zinc, which are carcinogenic.[13,14] Finally, elements like iron can form compounds that may suffocate fish and bottom-dwelling aquatic life.[15]

3) *Suspended solids* in preparation plant wastewater and runoff, if not properly contained, can constitute a third source of pollution in surface-water systems.[16] These solids, containing fine coal and clay particles and often trace elements,[17] are usually created when rainwater erodes the surface of waste piles. Suspended solids may also be released in the overflow from slurry ponds where solids have not adequately settled.[18] They can fill local streams with silt, discouraging the hatching of fish eggs and polluting drinking-water sources. Further, siltation of shipping channels increases the need for dredging.[19]

4) Fine *dust* particles in refuse piles and along roads near waste-disposal facilities can become airborne on windy days, creating air pollution around disposal sites. Some slurry ponds have areas of exposed fine refuse above the water line, which can also become sources of airborne particulates.[20]

5) *Noxious gases* released from burning waste piles are the last potential source of pollution. These gases may contain hydrogen sulfide, sulfur oxides, nitrogen oxides, carbon monoxide, hydrocarbons, POMs (polycyclic organic matter), and trace elements such as arsenic, boron and mercury.[21]

Fires can start in waste piles from spontaneous combustion of wastes, the intentional burning by plant personnel of coal wastes and trash near waste piles, and accidental brush or forest fires. Spontaneous combustion is a particularly serious hazard. It occurs when the oxygen present in the air and the organic matter and/or pyrites present in the wastes react to generate enough heat to start a fire. In general, waste from lower-ranked coals such as subbituminous and lignite is more likely to ignite than waste from higher-ranked coals such as anthracite and bituminous.[22]

Characteristics of Preparation Plant Wastes

The physical characteristics of preparation plant wastes influence the effect they will have on the environment. These wastes are composed primarily of soft clay and shale particles that crumble easily into smaller-size fractions. They may also include sandstone and limestone particles. The size fractions of the wastes vary, depending mainly on the properties of the raw coal, the kind of rock found around the coal seam, the method of mining, the individual preparation plant procedures, the methods of handling and storing wastes, their age, and the amount of compaction they undergo.[23]

Wastes are commonly described as either "coarse" or "fine." Coarse wastes are made up of sand and of gravel-sized particles that are larger than one millimeter at their widest point, while fine wastes are the clay, silt and smaller sand particles that are less than one millimeter at their widest point.[24]

Fine refuse contains a relatively high percentage of moisture (8 to 56 percent), as well as a 30 to 60 percent ash content, making it susceptible to degradation and not very useful as a construction material.[25,26] Coarse refuse, on the other hand, has a much lower moisture content (2 to 28 percent), and an ash content of 55 to 75 percent.[27,28] Coarse refuse has been used as a construction material in both dams and roadway bases in the United States, Britain, Germany and the Netherlands.[29]

Preparation plant wastes differ also in their permeability (the ease with which a liquid such as water can pass through them). Permeability is directly related to the presence of spaces between individual particles of wastes. Thus, coarse wastes are highly permeable, but their permeability can be reduced by mixing them with fine refuse that fills in the spaces. Because fine refuse is relatively impermeable, it makes a better bottom liner for impoundments or settling ponds.[30]

Permeability has an important effect on spontaneous combustion as well. Highly permeable wastes allow oxygen to penetrate the waste pile, and contribute to the likelihood of its heating up.[31]

Methods of Disposal

Nearly all the environmental problems linked to preparation plant waste disposal, excluding waste-pile fires and airborne particulates, are due directly or indirectly to the leaching of water through the waste material into its surroundings. These problems are caused by (1) locating waste-disposal sites in the path of water runoff and in the vicinity of sensitive plant and animal life; (2) the improper grading and reclamation of waste-disposal sites; or (3) an improper mix of waste-material sizes in disposal sites, which leaves the wastes susceptible to degradation and allows water to leach through. Nevertheless, damage to the environment can be limited by refuse piles, slurry ponds, and underground burial sites that are properly located, constructed and operated.[32.]

Refuse Piles/Landfills: Refuse piles and landfills are the most common means of preparation plant waste disposal. Eight of the 12 preparation plants profiled by INFORM reported that they disposed of at least part of their wastes in this manner. Although the terms "refuse pile" and "landfill" are often used interchangeably, landfills usually refer to the more sophisticated surface-burial sites, where wastes are spread, compacted, and covered regularly.

In the past, refuse piles were often placed in locations that were the most easily accessible by the cheapest available modes of transportation. For example, if an aerial tramway was the easiest and least expensive way to transport waste at a given plant, the wastes were simply dumped from the trams into piles or valley impoundments without consideration of environmental effects.[33]

Since the passage of the major federal environmental protection laws of the 1970's, more care has been given to the choice of refuse sites and to their design, construction and operation. The type of refuse (coarse or fine), the topography of the site, the mode of transportation, and the regulatory requirements now enter into the siting and design decisions about waste piles.[34]

There are several recommended procedures for constructing and operating waste piles. EPA advises that if water seepage or springs are present at the site, a properly constructed refuse pile will have subdrains built beneath it to collect infiltrating water. (This reduces the potential for water pollution and improves the stability of the refuse pile.) Additionally, runoff will be diverted around piles via ditches. Both coarse and fine dewatered will be combined in waste piles to decrease their permeability.[35]

In general, preparation plant wastes are spread in layers and then compacted. This process keeps the waste pile stable, reduces oxygen exposure—minimizing the risk of spontaneous combustion—and decreases the likelihood of water seepage into the pile.[36]

The exposed surfaces of waste piles, according to EPA, should be reclaimed by covering the inactive or completed sections of the pile with soil and then planting vegetation. The surface should be carefully sloped to avoid erosion and the "ponding" of waters on the pile.[37]

Slurry Ponds: If properly constructed and maintained, slurry ponds are an acceptable means of fine waste disposal.[38] One of the 12 preparation plants profiled by INFORM reported that it used slurry ponds to deal with its fine wastes.

Under proper conditions, fine wastes are submerged in water, where oxidation of their pyritic sulfur will be prevented and the formation of acid byproducts thus reduced.[39]

Slurry ponds are either "open" or "closed" systems. In an open system, clarified waste water can be discharged into ponds, lakes or streams. However, the potential for water pollution in an open system is significant, since settled solid materials can easily become resuspended if storm or windy periods occur.[40]

Regulations promulgated under the Clean Water Act set limits for discharges from preparation plants, limits which encourage the recycling and reuse of plant water. These limitations require slurry ponds to be closed systems, except under specific conditions when plants may discharge minimal amounts of pollutants into the environment.[41,42]

A closed system reduces the possibility of surface water pollution by recirculating most of the preparation plant water, after clarification, back into the plant. Unfortunately, the use of closed systems can create in-plant engineering difficulties, as corrosive salts tend to build up in the recirculated water.[43]

The limits on total suspended solids discharged into natural water supplies by preparation plant slurry ponds are set forth in Section 301 of the Clean Water Act. These guidelines vary, depending on whether preparation plant discharges are acidic or alkaline. Discharge limitations can be exceeded in two cases: (1) If the waste facility is designed to handle a ten-year, 24-hour precipitation event,* excess discharges resulting from a larger precipitation event are exempted; and (2) If, after neutralization and sedimentation, the resulting discharge does not comply with limitations on the element manganese, the alkalinity of the effluent may exceed limits to a "small" extent.[44]

EPA notes that in maintaining a slurry pond, it is important to allow adequate time for the suspended solids to settle. In addition, the water should be kept deep enough to minimize exposure of waste particles to the air, as the particles could become airborne under windy conditions. The impounding structure of the pond should be built of stable materials, to avoid acid leaching and structural dangers. To properly reclaim a slurry pond that is filled, the remaining water should be drained, and soil and vegetative cover should be reestablished.[45,46]

Underground Disposal: Although underground disposal of preparation plant wastes is a common practice in Europe, it has not been popular in the United States, where less expensive waste piles and slurry ponds are used more often. In the U.S., underground waste disposal has been used mainly to prevent the collapse of abandoned mines and the resulting subsidence of the land above the mines.[47,48]

Underground disposal frees surface land for more useful purposes, and it avoids surface pollution problems.[49] However, its benefits must be weighed against its higher cost, the increased dust, fire, and noise hazards encountered during disposal as compared with surface disposal, and the possibility of ground-water pollution from acid drainage. Furthermore, in order to control land subsidence over the mine sites, the wastes must include a high percentage of coarse particles. According to Bechtel Corporation, underground burial will probably not become widespread in the United States unless regulations governing surface waste disposal become more stringent, limiting the latter.[50,51]

---

*A ten-year, 24-hour precipitation event is the maximum amount of precipitation within a ten-year interval that would accumulate in 24 hours.

Treatment of Preparation Plant Wastes

Los Alamos Scientific Laboratory (LASL) evaluated four major strategies to reduce the environmental impact of preparation plant wastes at three stages: (1) before disposal, (2) at the time of disposal, and (3) after disposal.[52]

The most promising pre-disposal technology involves grinding the wastes to a very small size (less than 20 mesh) and heating the material to 600 °C to 1,200 °C in the presence of air for several hours. This process, called "refuse calcination," produces a harmless glasslike material that is similar to the "bottom-ash" in coal-fired power plants. The calcinated refuse, when buried in conventional landfills, will not release acid runoff, and potentially harmful trace elements are immobilized in the structure of the material.[53,54,55] However, the cost of refuse calcination is high relative to other waste-disposal technologies; it ranges from $3.40 to $9.89 per ton of cleaned coal.[56] This is due in part to the need for flue gas scrubbing during calcination, as 95 to 100 percent of the sulfur in the wastes is driven off in the form of sulfur dioxide.[57] According to LASL, this cost can be offset by the recovery of sulfur and iron by-products in the calcinated wastes, assuming that the price of the recoverable materials is sufficiently competitive.[58]

A recent report by LASL suggests that wastes can also be successfully pretreated by grinding and slurrying them with a lime/limestone mixture to neutralize some of the most toxic pollutants. Preliminary results of experiments with the slurried refuse indicate that it contains low concentrations of trace elements and reduced levels of acid leachates.[59] An economic analysis of this technology claims that the cost of the process ranges from $0.22 to $0.50 per ton of cleaned coal[60]—which would make this one of the least expensive environmental-protection technologies.

At the time of disposal, preparation plant wastes may be mixed with lime, limestone, alkaline soil, or waste material such as fly ash. These substances can neutralize the acid runoff and reduce the amount of trace elements released from the wastes.[61,62,63] The cost of this practice when alkaline soil is used is described by LASL as "moderate," ranging from $0.57 to $1.69 per ton of cleaned coal.[64] However, questions remain about the long-term effectiveness or permanence of this technique.[65,66]

A related treatment surrounds waste particles with cement, clay or soil. These surface coatings isolate sulfur wastes from air and water, reducing the formation of acids and the release of trace elements.[67]

Acid drainage that occurs after waste disposal can be collected and treated several ways. LASL found "alkaline neutralization" the most promising post-disposal technology. Here lime, limestone or lye is added to the waste drainage to decrease its acidity and inhibit the release of trace elements. This procedure is often used to treat acid drainage from coal mines and therefore has the advantage of being familiar to the mining industry.[68,69,70] Although the costs of the treatment are low, ranging from $0.10 to $0.80 per ton of cleaned coal,[71] it has several important disadvantages. First, since the source of the acid drainage (e.g., the waste pile) is not treated, alkaline neutralization must be kept up indefinitely. Also, acid leachates from refuse disposal sites can escape before treatment.[72,73]

The following table summarizes the environmental-protection options discussed in the preceding section:[74,75,76]

	Before Disposal	At Time of Disposal	After Disposal
	Calcination	Lime/Limestone Slurry	Alkaline Soil "Co-disposal"	Alkaline Neutralization
Cost (per ton of cleaned coal):	$3.40 to $9.89	$0.22 to $0.50	$0.57 to $1.69	$0.10 to $0.8
Effectiveness:	Good	Good	Good	Good
Complexity:	High	Moderate	Moderate	Moderate
Treatment Duration:	Short	Short	Short	Long
Permanence:	Good	Uncertain	Uncertain	Poor

Use of Preparation Plant Wastes

Coal preparation plant wastes have been used experimentally as construction materials for roads, dams and landfills. For example, in Ebensburg, Pennsylvania, coal wastes are being employed for paving a highway and for landfill at the site of a prospective industrial park.[77] West Virginia as well as Pennsylvania has successfully used preparation plan wastes as road-paving material.[78] R&F Coal Company is building a dam out of the coarse wastes from its preparation plant in Warnock, Ohio (see R&F profile). Other construction-related products of preparation plant wastes include anti-skid materials, lightweight aggregate (cement filler), and conventional bricks. However, some questions remain about long-term durability of such wastes for these purposes.[79,80] Further, the acid drainage from plant wastes remains an environmental threat when they are used in construction. As in conventional waste disposal, the refuse may create acid runoff and contaminate surface and groundwater systems if improperly handled.

Preparation plant wastes can be sources of fuel and/or metals. If the Btu content of coal wastes found in slurry ponds and waste piles is sufficiently high, the refuse may be reclaimed as an energy source.[81] A research group at the University of Alabama tested 40 slurry ponds in Alabama. They found that the heating value of the wastes went up to 10,000 Btu per pound. Those in the upper range of heating value could be reclaimed as fuel.[82] In addition, Oak Ridge National Laboratory is investigating methods of recovering aluminum, iron and titanium from coal wastes.[83]

Cost of Waste Disposal

The popularity of refuse piles/landfills and slurry ponds, in which the bulk of U.S. coal preparation plant wastes are treated, is due largely to their low cost when compared to alternatives such as underground disposal and the reuse of wastes as construction materials.

Law Engineering Testing Company of McLean, Virginia, did a detailed cost analysis of three commonly used disposal techniques, employing waste piles and slurry ponds, for both coarse and fine preparation plant wastes. The following table lists the results of this study:[84]

METHODS AND COSTS OF WASTE DISPOSAL

Waste Disposal Method	Cost per Ton* of Refuse	Cost per Ton of Cleaned Coal
Compacted coarse refuse used to build an impoundment containing fine-waste slurry	$1.54	$0.66
Mechanical dewatering of fine refuse; combination of dewatered fine refuse with coarse wastes in a refuse pile	$1.73	$0.74
Slurry ponds used to separate fine refuse from slurry water; fine refuse dredged from slurry pond and combined with coarse waste in refuse pile	$1.92	$0.83

*(1979 dollars)

The preparation plant used as a basis for this study processed 1,000 tons of raw coal per hour, generated 300 tons of wastes per hour, and employed heavy-media cyclones to clean the coal. Law Engineering concluded that its cost analysis was compatible with data supplied by American Electric Power-owned preparation plants; refuse-disposal costs averaged between $1.50 and $2.00 per ton of waste both in Law Engineering's hypothetical analysis and in actual plant performance.[85]

Duquesne Light Company and Northwest Coal Corporation were the only preparation plant owners profiled by INFORM that provided waste-disposal cost figures. (The others treated the figures as proprietary information.) In 1975 the coarse and fine refuse from Duquesne's preparation plant was disposed of in compacted waste piles at a cost of $0.76 per ton of cleaned coal, a figure that agrees with Law Engineering's data. The cost rose to $1.47 in 1979. The utility did not explain this increase; however, greater amounts of ash and mining wastes have been extracted from the raw coal cleaned by Duquesne in recent years, which may help account for the higher cost.*

Bechtel Corporation and other researchers have estimated that the cost of underground waste disposal would be over a dollar more per ton of cleaned coal than the cost of surface disposal for the same amount of waste. The cost of using preparation plant wastes as construction materials has been reported by Bechtel as being "relatively high," due primarily to the expense of transporting the wastes to locations where they can be used.[86]

Regulations Controlling Preparation Plant Wastes

The laws regulating pollutants produced by preparation plants and other sources of mining wastes were enacted in response to some serious problems in the past. The 1972 Buffalo Creek Valley flood, which resulted from the collapse of a Pittston Company coal-waste embankment in Logan County, West Virginia, killed 125 persons and caused millions of dollars in property damage. This disaster in particular prompted a federal study of the dangers of coal-waste impoundments and resulted in stronger federal and state laws and regulations governing waste-pile and impoundment stability.[87] Also, increased public awareness of the harm done by acid drainage has led to a gradual strengthening of laws, both federal and state, controlling such discharges from mines and preparation plants.[88]

Some of these measures—which are discussed below—overlap, presenting problems of redundancy and inconsistency when applied at the same time. The federal agencies and Congress, recognizing the need to simplify this legislation, have taken steps to integrate or coordinate its multiple requirements. In addition, as a result of coal industry appeals of lower court rulings, certain aspects of the laws (for example, some of the information required for Surface Mining Control and Reclamation Act [SMCRA] permits) are scheduled for review by the Supreme Court in 1981.[89]

SMCRA is the most comprehensive law regulating pollution from preparation plants, with sections addressing both air and water pollution and solid-waste control. SMCRA was enacted to protect the quality of streams, runoff, and other surface and groundwater systems, and to minimize changes in their location and volume. The law sets conditions for process-waste disposal and performance standards for the construction of waste-impoundment areas, sedimentation ponds, waste banks, dams, and underground mine areas.[90]

---

*Northwest Coal reported that its waste-disposal cost was $3.00 per ton of refuse (1981 dollars). The company attributes this high cost to its shipping of wastes to a county landfill 26 miles from the preparation plant. It is now looking for land near the plant, where it plans to build a waste pile.

Aside from SMCRA, the primary federal statute protecting U.S. waters from pollution by coal cleaning plants is the Federal Water Pollution Control Act, known as the Clean Water Act (CWA). Through a system of permits issued by the EPA or EPA-approved state programs, the CWA limits the discharge of the acid or alkaline effluents, suspended solids, trace elements, and toxic or hazardous substances that are found in plant runoff, spill waters and waste disposal sites. The Safe Drinking Water Act (SDWA) regulates wells where wastes are injected into the ground at preparation plant sites, since this practice of waste disposal may directly affect groundwater quality.[91]

Regulations promulgated under both SMCRA and the Resource Conservation and Recovery Act (RCRA) can potentially affect management of the solid wastes produced by preparation plants. However, controls for such solid wastes are now found only in the regulations under SMCRA. RCRA-mandated regulations may be expected in the future if EPA classifies preparation plant wastes as hazardous.*

At present, the Toxic Substances Control Act (TSCA) does not directly affect coal preparation plant operations. However, chemicals not yet developed, such as flocculants and other reagents used in chemical coal cleaning, may be regulated by this statute in the future.[92]

The Clean Air Act (CAA) controls air pollution generated by preparation plants and waste-disposal areas. Regulations cover particulate emissions from thermal dryers and pneumatic cleaning equipment, as well as fugitive dust emissions from haul roads, conveyors, crushing operations, disturbed land, and coal-storage or waste piles. The CAA is being reviewed in 1981, and it is not yet clear what changes will result. SMCRA regulations, in addition to those of the CAA, deal with fugitive dust emissions.[93]

The federal Mine Safety and Health Act (MSHA) covers worker and citizen issues—such as respirable-coal-dust limitations and the stability and safety of waste piles, landfills, dams and embankments—that relate to coal preparation plants. Also, noise-control standards for plant equipment have been established under MSHA.[94]

The following table lists the major federal laws that apply to coal-preparation activity, the agencies that administer them, and the laws' areas of jurisdiction:[95]

Environmental Laws That Apply To Coal Preparation

Law	Administering Agency	Major Areas of Jurisdiction
Surface Mining Control and Reclamation Act	U.S. Department of the Interior: Office of Surface Mining	Surface and ground-water quality, waste impoundments, dams, solid-waste disposal, fugitive-dust emissions

---

*A spokesman for EPA indicated that preparation plant wastes will probably not be classified as hazardous, given the current "anti-regulatory" climate in the federal government.

Clean Water Act	U.S. EPA: Office of Water and Waste Management	Pollutant-discharge permits, water quality, hazardous or toxic pollutants
Safe Drinking Water Act	EPA: Office of Water and Waste Management	Waste-injection wells (that may affect groundwater quality)
Resource Conservation and Recovery Act	EPA: Office of Water and Waste Management	Hazardous-waste permits
Toxic Substances Control Act	EPA: Office of Pesticides and Toxic Substances	Hazardous or toxic chemicals
Clean Air Act	EPA: Office of Air, Noise and Radiation	Particulate pollution, fugitive-dust emissions, SO_2 from thermal dryers
Mine Safety and Health Act	U.S. Department of Labor: Mine Safety and Health Administration	Waste-pile, dam, embankment stability; respirable dusts; noise

Amenability of Coal to Preparation

U.S. coals differ greatly in composition. Their sulfur, ash, moisture and trace element concentrations vary according to the regions, mines and even individual coal seams in which they are found. Raw coal is cleaned to the degree specified by the individual coal users, and their specifications, which include allowances for maximum sulfur and moisture, and minimum ash and Btu content, are in turn determined by the economic and regulatory requirements affecting each user. The cost of cleaning coal to a specified level also depends on the relative concentration of impurities within a specific coal sample.[1]

A "washability analysis" determines the concentration of a coal's impurities, and the type and cost of the potential cleaning process used to separate them from the coal.[2,3,4] This analysis is a laboratory procedure that approximates the operating conditions found in an individual commercial coal cleaning plant. It is performed by preparation plant engineers before a plant is designed, to help the designer determine what equipment and operating specific gravities are necessary, and to monitor the amenability of the raw coal feed to the cleaning processes employed during normal plant operations.[5]

The "float-and-sink" test comprises the heart of the washability analysis. It provides data on the specific gravity of the cleaned and refuse portions of the coal sample, and the specific gravity of the medium which must be used in the preparation plant equipment. In this test, raw coal samples are placed in a series of baths, each with a different specific gravity, to determine which specific gravity is best suited to separate that coal from its impurities.[6] Equally important is a test to determine the optimal size at which coal can be cleaned to the desired sulfur and ash specifications. Here, various size fractions of the raw coal sample are prepared to determine the extent to which the coal's impurities are liberated as the coal is crushed into smaller and smaller particles.[7] Since problems with handling, water removal and other factors that affect cleaning costs increase as particle size decreases, it is important to reach a balance between the variables of coal particle size, specific gravity, and the *desired* ash and sulfur levels specified by the user.[8] The relationships between these variables are defined by washability curves which record the results of the float-and-sink test.[9,10] Generally speaking, the cleaner the coal, the higher the costs incurred in its preparation.

Regional Amenability

The coal found in certain regions of the United States is more amenable to cleaning than coal found in others. Most of that cleaned in this country is eastern and midwestern bituminous coal (although some bituminous and sub-bituminous coals found in the western states are cleaned).[11,12]

Due to their high moisture content (up to 50 percent), western sub-bituminous and lignite coals are poor candidates for coal cleaning, as wet physical coal cleaning could contribute additional moisture and offset any benefits derived from sulfur and ash reduction. In addition, sub-bituminous and lignite coals usually contain only small quantities of sulfur and ash. At the present time, most western sub-bituminous and lignite coals are only crushed for size control before shipping. However, if western sub-bituminous coals contain large amounts of mining wastes, they are subjected to minimal preparation (to level 2).[13]

Amenability of Coal to Sulfur Reduction

Regulations governing sulfur dioxide emissions have introduced a new factor in determining the degree to which coal is cleaned. With the passage of the Clean Air Act of 1970, mandating sulfur dioxide emission control, increased attention has been focused within the industry on the effectiveness of different levels of coal cleaning in removing sulfur from different types of coal. As a result, many studies have set out to determine what U.S. coals can potentially be cleaned to the degree necessary for compliance with the law.

The U.S. Bureau of Mines analyzed 445 raw coal samples representing every region of the country, to determine the potential uses of physical coal cleaning processes for removing sulfur from raw coal.[14] The regional results of this analysis are presented in Tables A and B. Table A lists the regional average of organic, inorganic and total sulfur. Table B lists the regional average of sulfur and ash that can be removed from raw coal samples by physical coal cleaning, as well as the percent of original Btus not lost during cleaning, and the calorific content and sulfur dioxide emissions of the cleaned coal product.

Since physical coal cleaning processes cannot remove organic sulfur from raw coals, the percent content of inorganic sulfur establishes an optimal level of sulfur reduction that could be obtained by physical coal cleaning processes. This optimal percentage of sulfur reduction would range from 34 percent of the original sulfur content of western coals to 68 percent of the original sulfur content of eastern and midwestern coals. In reality, this optimal level cannot be obtained, and the range given above is lowered to a 15 percent and 44 percent (respectively) reduction in the original sulfur content of the raw coal samples. The actual range of these figures is due in large part to the fact that commercial coal cleaning plants are not always capable of removing fine-sized inorganic sulfur particles from the raw coal feed.[15]

More specifically, northern Appalachian and midwestern coals contain a high percentage of inorganic sulfur, some of which can be removed by physical coal cleaning. However, the reduction in organic sulfur is generally not great enough to meet the 1971 New Source Performance Standard (NSPS) of 1.2 lb $SO_2/10^6$ Btu. In contrast, many coals from southern Appalachia, Alabama and the western regions contain less sulfur and can meet the 1971 NSPS either raw or cleaned.[16,17]

The U.S. Bureau of Mines report concluded that only 14 percent of the total U.S. coal samples studied could meet the 1971 NSPS without cleaning. If the coal samples were crushed to 14 mesh (a very small coal size) and 50 percent of the raw coal Btu value was lost in reject materials, 32 percent of the coal samples could meet the 1971 NSPS.[18] It is important to note that the Revised New Source Performance Standards (RNSPS) were not considered in this study. They are considerably more stringent than earlier standards, and physical coal cleaning alone is not capable of removing enough sulfur from raw coal to meet the environmental standards set for sulfur dioxide emissions.

In 1980, Teknekron Research, Inc. performed a similar study on samples of U.S. coal reserves.[19] Coals from each region of the country were analyzed for sulfur content, potential sulfur dioxide emissions, and amenability to physical coal cleaning. In addition, five other technologies were considered (chemical coal cleaning, fluidized-bed combustion, flue gas desulfurization, and low and medium Btu gasification) as a means of reducing sulfur dioxide emissions (Table C lists some of the results of this report). Each regional coal was examined to determine whether it could meet the sulfur dioxide emission limitations of 1.2, 2.0 and 3.0 lb $SO_2/10^6$ Btu. According to Teknekron, 38 percent of all U.S. coal reserves can meet a 1.2 lb $SO_2/10^6$ Btu emission standard, as illustrated in Table C, in the raw form. Physical coal cleaning (1.5″ top size, 1.6 specific gravity) can increase this figure to 50 percent. Table C also illustrates the regional differences in coal sulfur content. Only 2 percent of all eastern and midwestern coals can meet the 1.2 lb $SO_2/10^6$ Btu emission standard when physically cleaned, as compared with 86 percent of all western coals.[20]

Table A: AVERAGE SULFUR VALUES FROM SIX US COAL REGIONS

Region	% Total Sulfur	% Inorganic Sulfur	% Organic Sulfur	Inorganic Total Sulfur(%)
Northern Appalachia	3.01	2.01	1.00	67
Southern Appalachia	1.04	0.37	0.67	36
Alabama	1.33	0.69	0.64	52
Eastern Midwest	3.92	2.29	1.63	58
Western Midwest	5.25	3.58	1.67	68
Western	0.68	0.23	0.45	34
Total US	3.02	1.91	1.11	63

Table B: Amenability of U.S. Coals to Sulfur Removal by Physical Coal Cleaning*

Region	No. of Samples	Btu Recovery (%)	Ash (%)	Inorganic Sulfur (%)	Total Sulfur (%)	Btu per lb+	Emissions of lbs $SO_2/10^6$ Btu+
Northern Appalachia	227	92.5	8.0	0.85	1.86	13,776	2.7
Southern Appalachia	35	96.1	5.1	0.19	0.91	14,197	1.3
Alabama	10	96.4	5.8	0.49	1.16	14,264	1.7
Eastern Midwest	95	94.9	7.5	1.03	2.74	13,138	4.2
Western Midwest	44	91.7	8.3	1.80	3.59	13,209	5.5
Western	44	97.6	6.3	0.10	0.56	12,779	0.9
Total US	445	93.8	7.5	0.85	2.00	13,530	3.0

*3/8 inch top size; 1.60 specific gravity
+moisture-free

Table C:
Percentage of Coal Reserves Able to Meet
Long-Term Average SO_2 Emission Limits

Regional Summary*

SO_2 Emission Limit (lb/MBtu)	Region	No. Control (%)	PCC+ (%)
1.2	Northern Appalachia	5	11
	Southern Appalachia	50	62
	Alabama	27	36
	Eastern Midwest	1	2
	Western Midwest	6	6
	Western	66	86
	TOTAL US	38	50
2.0	Northern Appalachia	15	30
	Southern Appalachia	83	88
	Alabama	68	75
	Eastern Midwest	6	9
	Western Midwest	11	13
	Western	90	96
	TOTAL US	56	62
3.0	Northern Appalachia	31	51
	Southern Appalachia	92	94
	Alabama	92	93
	Eastern Midwest	9	18
	Western Midwest	15	19
	Western	96	99
	TOTAL US	63	70

*Revisions same as previous tables
+Physical coal cleaning at 1½ inches, 1.6 specific gravity

Description of Coal

Coal is found in 31 states and mined in 26.[1] Approximately 13 percent of the land area of the United States contains deposits of varying size and quality. About two thirds of the land areas of Illinois and West Virginia, and more than 40 percent of Wyoming's and North Dakota's land areas, hold coal deposits.[2] Figure 1 illustrates the distribution of coal reserves in the United States (see page 339).

In general, each region of the country is known for a particular type of coal.[4] Table 1 lists the amounts of U.S. coal reserves, by region and by coal rank. (See Coal Ranking in this appendix.)

Coal has a physical and chemical composition that is highly variable. Different coals contain greatly differing percentages of combustible matter, noncombustible ash, sulfur, trace elements, and water. Coals also vary widely in the physical properties that affect how they are mined, transported, processed, stored, and burned as fuel.

The formation of coal is a complex process, involving the biological and geological transformation of decomposing plant matter over millions of years. The physical and chemical structure of coal is essentially a function of the extent to which the plant matter has been transformed into the substance we call "coal" and of the diverse biological, chemical and geological influences on that transformation.

The resulting differences in physical and chemical structure are used as the basis of a system of ranking coal types. Generally, the higher-rank coals are those that have, over geological time, been more completely transformed from the original plant materials.

US Coal Reserves by Region and Coal Rank*
January 1, 1976
(Million Tons)

COAL RANK

Region	Anthracite	Bituminous	Sub-Bi-tuminous	Lignite	Total	% of Total US Coal Reserves
East (Appalachia)	7,247	103,292	-	1,083	111,622	19.7
Mid-West	97	101,197	-	3,208	104,502	16.7
West	28	24,222	168,408	28,900	221,558	63.6
TOTAL	7,372	228,711	168,408	33,191	437,682	100.0

*Figures from *Coal Data Book: the President's Commission on Coal,* February, 1980

Coal Formation

According to most theories of coal formation, coal is formed from plant matter present along the fringes of lakes and in marshes and swamps. The basic organic building block of all plant matter is cellulose, which is composed of carbon, hydrogen and oxygen. Plant material also contains smaller concentrations of nitrogen, trace minerals, organic sulfur, and calcite.[5]

When these plants die and decompose, they fall into the water, become water-logged, and sink to the bottom. The plant matter continues to decompose in this environment of bacteria, fungi, and limited oxygen supply. The slow decay transforms the plant material into "peat," a substance with physical and chemical properties that are different from the original plant cellulose. Sand and clay impurities deposited from flood waters often become incorporated in the plant matter or peat. The presence or absence of water, oxygen, various minerals, and bacterial activity determines the structural form that the peat will take. These biological, chemical and geophysical processes will have an impact on the eventual form of the coal.[6]

Over many centuries, peat deposits build up to the water level and a "peat moss" grows on the exposed surface. As the peat becomes firmer, trees and other land vegetation grow there. The death and decay of this vegetation add to the volume of the decomposing plant material in the peat bed. The formation of a peat bed is the last stage in the life cycle of the original lakes, marshes and swamps. Throughout this phase of peat formation, the surrounding mineral and chemical environment, as well as bacteria and fungi, continues to alter the physical and chemical structure of the peat in a way that will partially determine the characteristics of the coal end product.[7]

Gradually, peat beds are covered by layers of rock-forming sediments, and pressure and heat from surrounding geologic structures slowly transform the peat into coal. The amount of pressure and heat applied, and the length of time that it is applied, determine the type of coal that is formed.* Lower-rank coals such as lignite are formed from the heat and pressure of overlying sediments. Higher-rank coals such as bituminous have been subjected to greater pressures. Anthracites, the highest-rank coal, are formed by more extreme pressures and temperatures resulting from movements of the earth's crust on sedimentary layers covering the peat beds. The higher-rank coals are therefore much more dense than the lower-rank coals, often occupying up to ten times less volume than the original plant material.[8]

Most bituminous coals found in North America date back to the Mississippian and Pennsylvanian geologic ages of some 270 to 350 million years ago. Lower-rank lignites and "sub-bituminous" coals (see Coal Ranking), found mostly in the western United States, date from the Cretaceous and Tertiary geologic ages of about 135 million years ago.[9]

As coal is transformed from peat, the proportions of moisture, oxygen and volatile matter (burnable gases) decrease, while fixed carbon and sulfur, and frequently ash, increase. Thus the higher-rank coals can be distinguished from the lower-rank coals by their higher fixed-carbon and lower volatile-matter compositions (distinctions to be taken up further in Coal Composition as well as Coal Ranking).[10]

Owing to the different forces at work during the formation of coal, as well as the variability of the layers of rock surrounding the coal beds, the characteristics of coal change from one region to another. For example, coals exposed to ocean water often contain more sulfur than coals formed in freshwater environments. In fact, coals may vary significantly between mines in a given region, and even within one seam or bed, both vertically and horizontally. [11,12]

Coal Composition

When coal is to be prepared and used as a fuel, its various components play different roles:

1) The *fixed carbon* content of the coal is composed of the coal's carbon atoms, which, along with the volatile matter, form the burnable portion of the fuel. Fixed carbon is important in determining the calorific, or heating, value of coal. The higher-rank coals (anthracite, low-volatile bituminous) have a high percentage of fixed carbon and a low percentage of volatile matter. The lower-rank coals (high-volatile bituminous, sub-bituminous) contain a lower percentage of fixed carbon and a higher percentage of volatile matter.[13]

---

*However, the age of a given coal does not necessarily indicate the coal's rank, since greater pressure may cause a "younger" coal to be higher in rank than an "older" coal.

2) *Volatile matter* is that portion of the coal released as gas when the coal is heated to a temperature of about 900 °C in the absence of air. These gases include methane (CH_4) and other hydrocarbons, carbon monoxide (CO), and some carbon dioxide (CO_2). Volatile matter does not include the moisture that is given off as water vapor when coal is burned. The percentage of volatile matter in the coal has an impact on its heating value.[14,15]

3) *Ash*, the noncombustible residue remaining after coal is completely burned, is formed from a portion of the noncombustible minerals found in coal.[16]

Two types of ash are generally present in coal. *Intrinsic ash* is composed of particles of pyrites (FeS_2), calcite $(CaCO_3)$, gypsum $(CaSO_4 \cdot 2H_2O)$, kaolinite $(Al_2O_3 \cdot 2SiO_2 \cdot 2H_2O)$, and other materials that were present in the environment where the coal was originally formed. This ash is present in a widely diffuse pattern in the coal and cannot be removed by physical coal cleaning processes. Therefore the percentage of intrinsic ash (usually less than 2 percent of the coal) defines the minimum ash content of a coal, regardless of its level of cleaning.[17,18]

Extraneous ash is composed of shale, fireclay, rock, bone, gypsum and calcite, and is present in coal in the form of larger-size impurities, often in bands in the seams. Much of the extraneous ash can be removed by physical coal cleaning processes. Extraneous ash should not be confused with the rock, shale and other wastes that lie over or under the coal seam and are commonly mined with raw coal. The amount of these other wastes produced at a given location depends on the size and quality of the coal seams, and the precision with which the coal is removed. These mining wastes are easily removed in coal cleaning plants; they often comprise a substantial percentage of the total waste volume.[19,20]

4) *Sulfur* is an important constituent of coal because of its role as an environmental contaminant. During normal combustion, all of the sulfur in coal is converted into sulfur oxides (SO_x)—mostly to sulfur dioxide (SO_2), with a small amount being further oxidized to sulfur trioxide (SO_3).[21] Physical coal cleaning, flue gas desulfurization, and fluidized-bed combustion* are all technologies that can reduce this problem.

Three kinds of sulfur are found in raw coal. *Inorganic sulfate sulfur* usually makes up less than 0.05 percent of the raw coal; it is water-soluble and therefore easily removed by coal cleaning.[22]

Pyritic sulfur (inorganic sulfur) was deposited in coal during its formation, along with the minerals that compose coal ash. It is the form of coal sulfur most amenable to coal cleaning because it is significantly more dense than coal and is distributed throughout raw coal as discrete particles (ranging in size

---

*For discussions of flue gas desulfurization and fluidized-bed combustion technologies, see INFORM studies *The Scrubber Strategy* and *Fluidized-Bed Energy Technology*.

from one micron to 40 mm). The amount of pyritic sulfur in coals varies.[23,24,25]

Organic sulfur is actually part of coal's molecular structure, and cannot be removed by physical coal cleaning processes. Chemical coal cleaning technologies have demonstrated the ability to remove some of a coal's organic sulfur content; however, such processes have not yet become commercially usable. Therefore the percentage of organic sulfur defines the maximum potential desulfurization that can be obtained by the physical coal cleaning processes. This form of sulfur constitutes between 30 and 70 percent of the total sulfur content of coal.[26,27]

5) Small amounts of *trace elements* are also found in coal. These trace elements generally occur in concentrations proportional to their prevalence in the earth's crust. However, the percentage of each element found in individual coal samples varies widely with the unique geologic history of the samples. The non-volatile trace elements make up ash, whereas the volatile trace elements are released into the atmosphere during coal combustion. Some of these trace elements, such as arsenic and lead, are dangerous to human health and the environment.[28,29]

6) *Moisture* is present on the surface, and within the molecular structure, of coal. As the rank of a coal increases, its "inherent" moisture content (as well as its volatile matter) generally decreases. The percentage of moisture in a given coal has an impact on its heating value.[30]

Coal Ranking

Coal ranking describes the extent to which plant matter has been transformed from lignite to anthracite, the two ends of the coal spectrum. The two major criteria for determining the rank of coal are fixed-carbon content and heating value. The percentages of volatile matter and moisture, and the characteristics of agglomeration, consolidation, and weathering, are examined to a lesser extent.*[31,32] Listed below are the ranks of coal by class and group and the criteria that define each category. Figure 1 illustrates the relationship between moisture, volatile matter, fixed carbon, and rank for 12 types of coal.

1) *Anthracite*

    Meta-anthracite is rarely mined or used as a fuel.

    Anthracite is mined primarily in eastern Pennsylvania. It is a smokeless,

---

*Agglomeration is the tendency of coal to form a ball when heated. Consolidation is the tendency of coal to remain in a solid form when mined. Weathering is discussed under Physical Properties below.

odorless coal that is often used in small-scale applications such as home heating stoves.

Semi-anthracite is an easier-kindling, higher-Btu fuel than anthracite. Semi-anthracite coal is not found in significant quantities in the United States.[33]

2) *Bituminous*

Low- and medium-volatile bituminous coals are mined in surface and underground mines in the eastern and midwestern United States. Such coals are nearly smokeless if burned efficiently, with a proper draft and oxygen flow. The sulfur content of eastern bituminous coals has a wide range.

High-volatile bituminous coals have a somewhat lower heating value and a higher percentage of volatile matter than the higher-rank bituminous coals.[34]

3) *Sub-bituminous*

Most *sub-bituminous* coal is surface-mined in the western United States, where low-sulfur seams are commonly found (relative to midwestern and eastern coals).

4) *Lignitic*

Lignites are coals with a woodlike texture, which tend to remain consolidated when mined.

Brown coal, a subgroup of the lignites, tends to disintegrate when mined. This characteristic distinguishes the brown coal from the lignites.[35]

Physical Properties

The following is a list of the various physical properties commonly employed by producers and users to categorize different coals. These properties vary among coals from different regions and mines, and even from different seams of one mine.

1) The *calorific or heating value* of coal is commonly expressed in Btu's or British thermal units. One Btu is the quantity of heat required to raise the temperature of one pound of water by one degree Fahrenheit. The percentages of carbon, hydrogen, sulfur, and volatile components in coal determine its heating value. Although the percentage of fixed carbon increases with coal rank, heating value does not always increase with coal rank. The percentage of volatile matter also has an important impact on heating value. For example, anthracite coals, which have a high percentage of fixed carbon and a relatively low percentage of volatile matter, have a lower Btu value than some of the

Table A: CLASSIFICATION OF COAL BY RANK

Class	Group	% Fixed Carbon	% Volatile Matter	Btu/lb	% Moisture
ANTHRACITE	1. Meta-Anthracite	98+	-2	+14,000	2 to 5
	2. Anthracite	92 to 98	2 to 8	+14,000	2 to 5
	3. Semi-Anthracite	86 to 92	8 to 14	+14,000	2 to 5
BITUMINOUS	1. Bituminous-Low Volatile	78 to 86	14 to 22	15,000+	2 to 5
	2. Bituminous-Medium Volatile	69 to 78	22 to 31	15,000+	2 to 5
	3. Bituminous-A-High Volatile	-69	31+	14,000+	5 to 18
	4. Bituminous-B-High Volatile	-69	31+	13,000 to 14,000	5 to 18
	5. Bituminous-C-High Volatile	-69	31+	11,000 to 13,000	5 to 18
SUB-BITUMINOUS	1. Sub-Bituminous-A	37 to 42	30 to 40	11,000 to 13,000	18 to 30
	2. Sub-Bituminous-B	37 to 42	30 to 40	9,500 to 11,000	18 to 30
	3. Sub-Bituminous-C	37 to 42	30 to 40	8,300 to 9,500	18 to 30
LIGNITIC	1. Lignite	+30	+25	6,300 to 8,300	30+
	2. Brown Coal	+30	+25	-6,300	30+

higher-rank bituminous coals, which have more volatile matter and less fixed carbon than the anthracites. This is because the higher concentration of burnable gases makes the coals kindle better and burn more efficiently. This trend is illustrated in Figures 1 and 2.[36]

2) The *temperature at which ash fuses* is an important characteristic of coal. Briefly, a drop in furnace efficiency can result from the fused ash interfering with the free flow of air necessary for proper combustion. In addition, large deposits of fused ash can form and reduce the transfer of heat between the furnace and water tubes by coating the inside surfaces of the furnace.[37]

3) The *specific gravity* of coal expresses the relationship between the density of water and the density of the fuel, as described in the following formula: Specific Gravity (SG) = Density of coal and coal wastes (pounds per cubic foot) divided by Density of water (62.4 pounds per cubic foot). Specific gravity is an important concept because many physical coal cleaning technologies use the difference between the specific gravity of coal and of its wastes to clean the fuel.*

Most coals have a specific gravity between 1.12 and 1.70 (lignites are not included in this range), depending on rank and on moisture, ash and sulfur content.† In general, specific gravities increase with rank, from lignite to anthracite, although a bituminous coal may have a higher specific gravity than an anthracite due to a higher ash content.[39,40]

Finely ground magnetite (or iron ore) is commonly added to water in coal cleaning plants to increase its specific gravity. If the specific gravity of the water-and-magnitite mixture is between that of the clean coal and the wastes, the two substances can be separated. Other materials used to increase the specific gravity of water in coal preparation plants include sand and calcium chloride.[41]

4) *Friability* is the susceptibility of a coal to breaking apart during handling and transportation. Highly friable coals can break down into fine-size fragments, leading to a variety of problems. Fine coal particles generate more dust during mining, transportation, loading, and unloading, creating handling and maintenance difficulties. The greatly increased surface area of fine coal fragments makes them much more susceptible to oxidation (or burning), and finally, fine coal particles are more difficult to clean.[42]

---

*If a given coal has a density (or weight per volume) of 85 pounds per cubic foot, then its specific gravity would be 85/62.4, or 1.36. Other liquids have relative densities or specific gravities that are different from that of water. Gasoline, which is less dense than water, has a specific gravity of 0.73, while mercury, which is significantly heavier than water, has a specific gravity of 13.55.[38]

†Ash and other impurities have a wide range of specific gravities: pyritic sulfur, 4.6 to 5.2; gypsum, kaolinite and calcite (typical ash constituents), 2.3, 2.6 and 2.7; and sandstone, clay and shale (typical mining wastes), 2.6.

5) *Grindability* is the ease with which coal can be made small enough to be used as a pulverized fuel. Coals from the same bed may differ widely in grindability. Most utilities require coal suppliers to meet a grindability specification.[43]

6) *Weathering*, or slacking, is defined as the susceptibility of a coal to deterioration from the effects of the weather. Generally, weathering is a problem only when coal is transported and/or stored outdoors prior to cleaning or combustion. Exposure to moisture and sun may lead to weathering, which breaks the coal down into fine-size fragments, creating the problems associated with highly friable coals.[44]

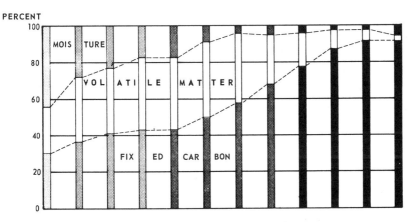

Figure 1. Moisture, Volatile Matter, and Fixed
Carbon by Rank (Moist, MMF Basis)
(courtesy of EPRI)

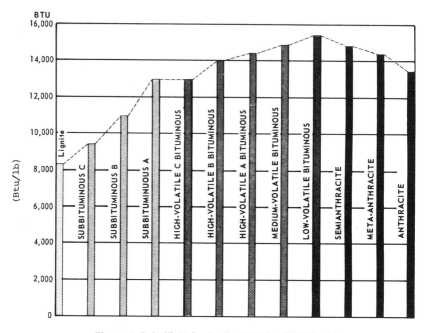

Figure 2. Calorific Value by Rank (Moist, MMF Basis)
(courtesy of EPRI)

United States Coal Deposits

The United States has been widely proclaimed "the Saudi Arabia of coal." By one account, enough U.S. coal deposits exist to supply energy at current rates of demand for more than 500 years.[1]

There are three ways of estimating the amount of coal found in the United States: (1) resources, (2) reserves, and (3) recoverable reserves.[2]

1) The term *resources* refers to all the coal deposits present at a given location, regardless of whether or not they can be mined. There are two subcategories of coal resources. "Total resources" refers to both the known and hypothetical unmined coal resources, whereas "resources" refers only to the identified coal deposits. Thus, coal resource figures are only a rough quantitative estimate. As of January 1, 1974, in the United States there were 3,968 billion tons of total (identified and hypothetical) coal resources to a depth of 6,000 feet, and 1,731 billion tons to a depth of 3,000 feet.[3,4]

2) *Coal reserves* are all the coal deposits that are considered potentially mineable. The thickness and accessibility of identified coal seams figure into the computation of reserves, and therefore there is less room for error than in the estimates of coal resources. There were an estimated 436 billion tons of unmined coal reserves in the United States as of January 1, 1974.[5,6]

3) *Recoverable reserves* are just those unmined coal deposits that can be removed using current mining technologies. The estimate of recoverable reserves takes into account the economic, legal, political and social variables that might prevent specific coal deposits from being mined. Recoverable-reserve estimates are revised when changes occur in the economic, technological, and other factors that limit or expand the recovery capabilities. Recoverable reserves is therefore the most accurate assessment of currently available coal deposits. The Office of Technology Assessment estimated that

as of January 1, 1974, there were 283 billion tons of recoverable reserves.[7]

In 1974 approximately 31 percent of the world's recoverable coal reserves were found in the United States; 23.1 percent were found in the U.S.S.R and 13.5 percent in the People's Republic of China. Other countries that compete at present with the United States in the world coal market include Australia, Canada, Poland and South Africa.[8]

Coal is also the most plentiful fuel in the United States in terms of Btu content. In 1979, of the known recoverable U.S. energy reserves (calculated by Btu content), 81.7 percent were from coal. Other fuels trailed coal significantly: bitumens and shale oil, 7.2 percent; uranium oxide, 4.2 percent; natural gas, 3.5 percent; petroleum, 2.8 percent; and natural gas liquids, 0.6 percent.[9]

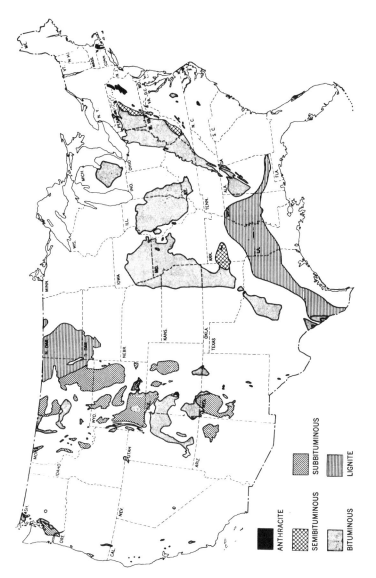

Coal Fields of the United States
(courtesy of EPRI)

ANTHRACITE

SEMIBITUMINOUS

BITUMINOUS

SUBBITUMINOUS

LIGNITE

Notes and Sources

Introduction
Notes

1. The President's Commission on Coal, *Coal Data Book* (Washington, DC: U.S. Government Printing Office, February 1980), p. 65.

2. U.S. Department of Energy, *Monthly Energy Review—August 1981 (Washington, DC: U.S. Government Printing Office, August 1981), p. 58.*

3. *The President's Commission on Coal, Coal Data Book,* p. 85. This figure, however, does not include anthracite coal, which in 1977 accounted for less than one percent of total coal production.

4. Ibid., p. 29.

5. Ibid., p. 35.

6. U.S. Department of Energy, *Monthly Energy Review,* August 1981, p. 64.

7. U.S. Environmental Protection Agency, *1977 National Emissions Report* (US Environmental Protection Agency: EPA-450/4-80-005, March 1980).

8. The National Academy of Sciences, *Atmosphere-Biosphere Interactions: Toward a Better Understanding of the Ecological Consequences of Fossil Fuel Combustion* (Washington, DC: National Academy Press, 1981).

9. John W. Green, Virgil Whetzel, Martha A. Petzel and Kenneth A. Ebeling, "An Assessment of the Capacity for Cleaning Coal in the United States" (Manuscript written by the Economics Impacts of Coal on

Natural Resources Project—U.S. Department of Agriculture/Economics Department, Colorado State University, September 1981), p. 6.

10.   *EPRI Journal,* "More Coal Per Ton," (Palo Alto, CA: Electric Power Research Institute, June 1979), pp. 6-7.

11.   John W. Green et. al., "An Assessment of the Capacity for Cleaning Coal," p. 7.

12.   Office of Technology Assessment, *The Direct Use of Coal* (Washington, DC: Office of Technology Assessment, April 1979), p. 76.

13.   Personal communication with James Kilgroe, Manager of Coal Cleaning, U.S. Environmental Protection Agency, 9/21/81.

14.   The President's Commission on Coal, *Coal Data Book,* p. 93.

15.   Office of Technology Assessment, *The Direct Use of Coal,* p. 76.

16.   The President's Commission on Coal, *Coal Data Book,* p. 85.

Findings
Notes

1.   James Cannon, "Why Electric Utilities Are Buying Coal," *Business and Society Review,* Winter 1980–81, Number 36, pp. 53-59.

2.   Personal communication with William Poundstone, Executive Vice President, Consolidation Coal Company, 11/6/81.

3.   U.S. Department of the Interior, Bureau of Mines, *Sulfur Reduction Potential of the Coals of the United States* (Washington, DC: U.S. Bureau of Mines, Report of Investigations/1976 RI: 8118).

4.   D.A. Sargent et al., *Effect of Physical Coal Cleaning on Sulfur Content and Variability* (U.S. Environmental Protection Agency: EPA-600/7-80-107, May 1980).

5.   Gibbs and Hill, Inc. *Coal Preparation for Combustion and Conversion* (Palo Alto, CA: Electric Power Research Institute, EPRI AF-791, Project 466-1, Final Report, May 1978) pp. 5-17.

6.   Hoffman Muntner Corporation, *Engineering/Economic Analysis of Coal Preparation with SO_2 Cleanup Processes* (U.S. Environmental Protection Agency: EPA-600/7-78-002, January 1978).

7.   James D. Kilgroe, "Combined Coal Cleaning and FGD," *Symposium on Flue Gas Desulfurization—Las Vegas, Nevada, March 1979; Volume I* (U.S. Environmental Protection Agency: EPA-600/7-79-167a, July 1979).

8.   PEDCo Environmental Inc., *Cost Benefits Associated with the Use of Physically Cleaned Coal* (U.S. Environmental Protection Agency: Contract No. 68-02-2603; Task No. 31, March 1980).

9. Gerald Blackmore, "The Outlook for Steam Coal as Used by Electric Utilities" (Paper presented to the American Mining Congress, 1980 Mining Convention, September 21-24, 1980, San Francisco), p. 15.

10. PEDCo Environmental Inc., *Cost Benefits Associated With the Use of Physically Cleaned Coal* (U.S. Environmental Protection Agency: Contract No. 68-02-2603; Task No. 31, March 1980).

11. Hoffman Muntner Corporation, *Engineering/Economic Analysis of Coal Preparation with SO$_2$ Cleanup Processes.*

12. James D. Kilgroe, "Combined Coal Cleaning and FGD."

13. E.F. Rubin and D.G. Nguyen, "Energy Requirements of a Limestone FGD System," *Journal of the Air Pollution Control Association,* Volume 18, No. 12, December 1978.

14. PEDCo Environmental Inc., *Cost Benefits.*

15. Tennessee Valley Authority/U.S. Environmental Protection Agency, *Evaluation of Physical/Chemical Coal Cleaning and Flue Gas Desulfurization* (U.S. Environmental Protection Agency: EPA-600/7-79-250, November 1979).

16. PEDCo. Environmental Inc., *Cost Benefits.*

17. U.S. Department of Commerce, "Sulfur Oxide Control Technology" (Photocopy; Washington, DC: U.S. Department of Commerce).

18. Gerald Blackmore, "The Utilities Perspective of Coal Utilization and Preparation," *Papers presented before the Symposium on Coal Preparation and Utilization, Coal Conference and Expo V, October 23-25, 1979, Kentucky Fair and Exposition Center, Louisville, KY* (NY: McGraw-Hill, Mining Information Services, 1979), p. 52.

19. Blackmore, "The Outlook for Steam Coal," p. 15.

20. Ibid., p. 16.

21. Ibid., p. 17.

22. Ibid.

23. Blackmore, "The Utilities Perspective," p. 54.

24. Blackmore, "The Outlook for Steam Coal."

25. Ibid., p. 23

26. Ibid., p. 24

27. M.K. Buder et. al., "The Effects of Coal Cleaning on Power Generation Economics" (Paper presented at the American Power Conference, Chicago, IL, April 23-25, 1979).

28. PEDCo Environmental, Inc., *Cost Benefits.*

29. Richard E. Briggs, "Rail Coal Rates: Past, Present and Future" (Paper

presented at the Coal Transportation Conference sponsored by Coal Week and Energy Bureau Inc., October 1980).

30. T.C. Campbell, "Transportation of a Premium Coal fines Product," *Fossil Energy* (U.S. Department of Energy, Process Evaluation Office, July 1979).

31. Michael M. Donahue, "The Western Railroads and Coal" (Paper presented at the Coal Transportation Conference sponsored by Coal Week and Energy Bureau, Inc., October 1980).

32. Gibbs and Hill, Inc. *Coal Preparation for Combustion and Conversion*, pp. 5-17.

33. Ibid.

34. *Business Week,* "Coal's Rough Ride Out of Appalachia," March 19, 1979, p. 148B.

35. Agis Salpukas, "Coal Boom Clogging Ports," *New York Times,* April 17, 1981, pp. D1, D6.

36. Bechtel National, Inc., *Impact of Coal Cleaning on the Cost of New Coal-Fired Power Generation,* M.K. Budar, Project Manager; (prepared for Electric Power Research Institute, Palo Alto, CA, EPRI CS-1622, Project 1180-2, Final Report, March 1981).

Methods of Coal Preparation
Notes

1. Gibbs and Hill, Inc., *Coal Preparation for Combustion and Conversion* (Palo Alto, CA: Electric Power Research Institute, EPRI AF-791, Project 466-1, Final Report, May 1978), pp. 2-1 to 2-3.

2. Bechtel Corporation, *Environmental Control Implications of Generating Electric Power from Coal, 1977 Technology Status Report, Appendix A (Part 1) Coal Preparation and Cleaning Assessment Study* (Prepared for Argonne National Laboratory; Work sponsored by the Division of Environmental Control Technology, U.S. Department of Energy, ANL/ECT-3, December 1977), pp. 222-224.

3. Gibbs and Hill, Inc., *Coal Preparation for Combustion and Conversion*, p. 2-45.

4. *EPRI Journal,* "More Coal Per Ton" (Palo Alto, CA: Electric Power Research Institute, June 1979), p. 10.

5. Ibid.

6. Howard W. Decker, Jr. and John N. Hoffman, *Coal Preparation Volume I* (University Park, PA: The Pennsylvania State University, 1963), p. 181.

7. Gibbs and Hill, Inc., *Coal Prepration for Combustion and Conversion,* pp. S-6 to S-8.

8. Decker and Hoffman, *Coal Preparation Volume I,* p. 167.

9. Ibid., p. 67.

10. Albert W. Deurbrouck and Richard Hucko, "Coal Cleaning and Desulfurization," *The Direct Use of Coal, Volume II–Part A: Working Papers, Appendix IV* (Washington DC: Office of Technology Assessment, April 1979), p. 25.

11. Ibid., p. 10.

12. Bechtel Corporation, *Environmental Control Implications of Generating Electric Power from Coal,* pp. 116-125.

13. Decker and Hoffman, *Coal Preparation Volume I,* p. 181, 192.

14. Gibbs and Hill, Inc., *Coal Preparation for Combustion and Conversion,* pp. 2-70 to 2-73.

15. Decker and Hoffman, *Coal Preparation Volume I,* p. 87.

16. Gibbs and Hill, Inc., *Coal Preparation for Combustion and Conversion,* p. 2-9.

17. Decker and Hoffman, *Coal Preparation Volume I,* p. 295.

18. Davis, Harold "PP&L Updates its Preparation Plant to Reduce Coal's Sulfur Content by 40%," *Coal Age Operating Handbook of Coal Preparation* (NY: McGraw Hill, 1978), p. 152.

19. Bechtel Corporation, *Environmental Control Implications of Generating Electric Power from Coal,* pp. 137-138.

20. Decker and Hoffman, *Coal Preparation Volume I,* p. 248.

21. Randhir Sehgal, Project Manager, Coal Cleaning, Electric Power Research Institute, Interview with INFORM, 10/81.

22. Bechtel Corporation, *Environmental Control Implications of Generating Electric Power from Coal,* pp. 141-142.

23. Decker and Hoffman, *Coal Preparation Volume I,* p. 253.

24. Bechtel Corporation, *Environmental Control Implications of Generating Electric Power from Coal* p. 135.

25. Gibbs and Hill, Inc. *Coal Preparation for Combustion and Conversion,* pp. 2-80, 2-81, 2-83.

26. Ibid., p. 2-80.

27. Bechtel Corporation, *Environmental Control Implications of Generating Electric Power from Coal,* p. 230.

28. Ibid., p. 133.

29. Gibbs and Hill, Inc., *Coal Preparation for Combustion and Conversion,* pp. 2-11 to 2-12.

30. Decker and Hoffman, *Coal Preparation Volume I,* p. 277.

31.   Gibbs and Hill, Inc., *Coal Preparation for Combustion and Conversion*, p. 2-88.

32.   Bechtel Corporation, *Environmental Control Implications of Generating Electric Power Coal*, pp. 149-150.

33.   Gibbs and Hill, Inc., *Coal Preparation for Combustion and Conversion*, p. 2-90.

34.   Decker and Hoffman, *Coal Preparation Volume II*, pp. 40-49.

35.   Bechtel Corporation, *Environmental Control Implications of Generating Electric Power from Coal*, pp. 151-152.

36.   P.W. Spaite, et. al., *Environmental Assessment of Coal Cleaning Processes: Technology Overview* (U.S. Environmental Protection Agency: EPA-600/7-79-073e September 1979), p. 33.

37.   Ibid.

38.   Bechtel Corporation, *Environmental Control Implications of Generating Electric Power Coal*, pp. 151-152.

39.   Gibbs and Hill, Inc., *Coal Preparation for Combustion and Conversion*, pp. 2-93 to 2-96.

40.   Decker & Hoffman, *Coal Preparation Volume II*, pp. 12-17.

41.   Duerbrouck and Hucko, "Coal Cleaning and Desulfurization," p. 19.

42.   National Coal Association, *Coal Data 1978* (Washington, DC: National Coal Association, 1980), p. II.

43.   Decker and Hoffman, *Coal Preparation Volume II*, pp. 52-60.

44.   Duerbrouck and Hucko, "Coal Cleaning and Desulfurization," p. 20.

45.   Gibbs and Hill, Inc., *Coal Preparation for Combustion and Conversion*, p. 2-45.

Introduction to Chemical Coal Cleaning
Notes

1.   U.S. Department of Energy, "Chemical Coal Cleaning" (Work in progress applying to fiscal year 1982).

2.   Personal communication with Albert G. Dietz, Jr., Program Manager, Coal Preparation, U.S. Department of Energy.

3.   Personal communication with Albert G. Dietz, Jr., and Richard Hucko, Supervisory Civil Engineer, Coal Preparation Division, U.S. Department of Energy.

4.   Personal communication with Albert G. Dietz, Jr., 10/30/81.

5.   R.A. Meyers, W.D. Hart and L.C. McClanathan, *The Gravimelt Process for Chemical Removal of Organic and Pyritic Sulfur from Coal* (Anaheim, CA: TRW, Inc., December 1979).

6.   Personal communicatin with Dr. Bernard Breen, Manager of Energy Technology and Projects, Reserach Cottrell, 7/24/80.

7. Bechtel Corporation, *An Analysis of Chemical Coal Cleaning Processes* (U.S. Department of Engery: Contract No. ET-78-C-01-3137, June 1980).

8. U.S. Department of the Interior, Bureau of Mines, *Sulfur Reduction Potential of the Coals of the United States* (Washington, DC: U.S. Bureau of Mines, Report of Investigations/1976 RI: 8118).

9. U.S. Department of Energy, "Chemical Coal CleaningTable A-1-"Technical Summary of Advanced Coal Cleaning Methods" with up-dated information provided by Albert G. Dietz, Jr.

10. U.S. Bureau of Mines, *Sulfur Reduction Potential of the Coals of the United States.* The Kentucky coal is form a Hopkins County coalbed; the Illinois coal is from Knox County. Both coals cleaned by heavy media separation with a 90% and 85% BTU recovery, respectively. Slightly better results may be available with froth flotation or oil agglomeration treatments.

11. TRW, Inc., unpublished experimental data. 1 = 1(NaK) OH fused salt treatment, 20 minutes, 370°C. Original sulfur levels 3.50% and 3.18% respectively.

12. U.S. Department of Energy, "Chemical Coal Cleaning," p. 5.

13. Richard Brown, ed., *Health and Environmental Effects of Coal Technologies* (McLean VA: The MITRE Corporation, Metrek Division, August 1979), pp. 90-145.

14. Personal communication with Dr. Peter Zavitsanos, Program Manager, General Electric Company, 7/24/80.

15. Personal communication with Dr. Thomas Wheelock, Professor, Iowa State University, 6/20/80.

16. Personal communication with Dr. George Snell, Manager, Gasification Projects, Hydrocarbon Research, Inc. 7/10/80.

17. Personal communication with Dr. Sidney Friedman, Chief of Coal Chemistry Branch, Pittsburgh Technology Center, 7/16/80.

18. Personal communication with Mr. John O. Siemsson, Project Manager, Atlantic Richfield Company, 8/80.

19. G.G. McGlamery, T.W. Tarkington and S.V. Tomlinson, "Economics and Energy Requirements of Sulfur Oxides Control Processes," *Proceedings: Symposium on Flue Gas Desulfrization—Las Vegas, Nevada March 1979; Volume I* (U.S. Environmental Protection Agency, EPA-600/7-79-167a, July 1979), pp. 201-214.

20. Pesonal communication with Dr. Scott Taylor, Supervisory Research Chemist, Pittsburgh Mining Technology Center, 7/7/80.

21. John A. Ruether, "Chemical Cleaning," *Combustion,* (NY: Combustion Publishing Company), December 1979, p. 25.

22. George R. DeVaux, "Kennicott Oxidative Desulfurization of Coal"

(Paper presented at the Second Conference on Air Quality Management in the Electric Power Industry, Austin, Texas, January 22-25, 1980), p. 8.

23. U.S. Department of Energy, "Chemical Coal Cleaning," p. 10.

24. Ibid., p. 5.

25. Ibid., p. 7.

26. G.Y. Contos, I.F. Frankel and L.C. McCandless, *Assessment of Coal Cleaning Technology: An Evaluation of Chemical Coal Cleaning Processes* (U.S. Environmental Protection Agency, EPA-600/7-78-173a, August 1978), p. 7, 95.

27. Howard W. Decker and John N. Hoffman, *Coal Preparation Volumes I and II* (University Park, PA: The Pennsylvania State University, 1963).

28. Bechtel Corporation, *An Analysis of Chemical Coal Cleaning Processes,* pp. 2-23 to 5-31.

29. Lee C. McCandless and Robert B. Shaver, *Assessment of Coal Cleaning Technology* (US Environmental Protection Agency, EPA-600/7-78-150, July 1978).

30. Robert A. Meyers, *Coal Desulfurization* (NY: Marcel Dekker, Inc., 1977), p. 55.

31. McCandless and Shaver, *Assessment of Coal Cleaning Technology.*

32. Bechtel Corporation, *An Analysis of Chemical Coal Cleaning Processes.*

33. V.C. Quackenbush, R.R. Maddocks and G.W. Higginson, "Chemical Communition—An Improved Route to Clean Coal," *Proceedings: Symposium on Coal Cleaning to Achieve Energy and Environmental Goals—September 1978, Hollywood, FL—Volume II* (U.S. Environmental Protection Agency, EPA-600/7-79-098b, April 1979).

34. Bechtel Corporation, *An Analysis of Chemical Coal Cleaning Processes.*

35. Ibid.

36. L.H. Beckberger et. al., *Preliminary Evaluation of Chemical Coal Cleaning by Promoted Oxydesulfurization* (Palo Alto, CA: Electric Power Research Institute, EPRI EM-1044, Project 833-1-April 1979).

37. Bechtel Corporation, *An Analysis of Chemical Coal Cleaning Processes.*

38. Sidney Friedman, "Preliminary Comments to a Panel Discussion on Prospects for Characterization and Removal of Organic Sulfur from Coal," *Proceedings: Symposium on Coal Cleaning to Achieve Energy and Environmental Goals—September 1978, Hollywood, FL, Volume II* (U.S. Environmental Protection Agency, EPA-600-/7-79-098b, April 1979), p. 1148.

39. Personal communication with Dr. Amir Attar, Professor, North Carolina State University, 7/3/80.

40. S. Friedman, R.B. Lacount and R.P. Warzinski, "Oxidative Desulfuri-
    zation of Coal," *Coal Desulfurization: Chemical and Physical Methods*
    (Washington, DC: American Chemical Society, 1977), p. 165.

41. Personal communication with Dr. Amir Attar, 7/3/80.

42. Personal communication with Dr. Scott Taylor.

43. Personal communication with Dr. Bernard Breen.

44. Personal communication with Dr. Peter Zavitsanos.

45. Personal communication with Dr. Amir Attar.

46. Personal communication with Dr. Bernard Breen.

47. U.S. Department of Energy, "Chemical Coal Cleaning," p. 12.

48. Ibid., p. 12.

49. Information obtained from each of the individual companies except for
    the percent of sulfur removed, percent of ash removed and percent of Btu
    loss which was provided by U.S. Department of Energy, "Chemical Coal
    Cleaning—Table A-1" with updated figures given by Albert G. Dietz, Jr.

Allen & Garcia Company
Sources

Allen & Garcia Company, "Services—Capabilities and Representative Proj-
    ects" (Chicago, Illinois: Allen & Garcia Company).

Brumbaugh, Owen E., Jr., Vice President of Engineering, Allen & Garcia
    Company. Interview with INFORM, 9/26/80.

Jackson, Dan, "High Grade Western Coal Goes East," *Coal Age,* June 1980,
    pp. 62-68.

Simon Engineering 1979 Annual Report (Great Britain).

Stock Exchange, London. 1979 Stock Exchange Official Yearbook, "Simon Engineer-
    ing" (East Grinstead, England: Thomas Skinner Directories).

Dravo Corporation
Sources

Dravo Corporation 1980 Annual Report and Form 10-K.

Moody's 1980 Industrial Manual, "Dravo Corporation," pp. 2111-2112.

West, Tom, Assistant to the Manager of Coal Sales and Development, Dravo
    Corporation. Interview with INFORM, 9/6/80.

Heyl & Patterson, Inc.
Sources

Griffin, Jack, Manager, Coal Group, Heyl & Patterson. Interview with INFORM, 11/12/80.

Heyl & Patterson, "Capabilities" (Promotional report published by Heyl & Patterson).

Notary, Joseph, Vice President, Process Division, Heyl & Patterson. Interview with INFORM, 11/12/80.

Kaiser Engineers, Inc.
Sources

Kaiser Engineers 1980 Annual Review.

Matoney, Joseph P., Manager, Coal Division, Kaiser Engineers, Inc. Interview with INFORM, 11/17/80.

Raymond International Inc. 1980 Annual Report and Form 10-K.

Ruppert, William J., Business Development Manager, Kaiser Engineers, Inc. Interview with INFORM, 11/17/80.

McNally Pittsburg Manufacturing Corporation
Sources

McNally Pittsburg, "McNally Coal Cleaning" (Promotional booklet published by McNally Pittsburg).

Draeger, Ernest, Chief Process Engineer, McNally Pittsburg. Interview with INFORM, 11/11/80.

Roberts & Schaefer Company
Sources

Elgin National Industries, Inc. 1980 Annual Report and Form 10-K.

Moody's 1980 Industrial Manual, "Elgin National Industries, Inc.," pp. 2128-2129.

Roberts & Schaefer Company, promotional pamphlets.

Wilson, John, Sales Contracting Engineer, Roberts & Schaefer Company, Interview with INFORM, 11/9/80.

A.T. Massey Coal Company
Sources

Coal Age, "A. T. Massey Joins Coal-to-Oil Venture," July 1980, p. 37.

Coal Age, "Improved Equipment Available Now," January, 1980, pp. 56-61.

Merritt, Paul C., ed., "Marrowbone Development Facility Now in West Virginia," *Coal Age Operating Handbook of Coal Preparation* (NY: McGraw Hill, 1978), pp. 206-207.

Moody's 1980 Public Utility Manual, "St. Joe Minerals Corp.," pp. 5153-5154.

St. Joe Minerals Corporation 1980 Annual Report and Form 10-K.

Salpukas, Agis, "The Transformation of St. Joe," *New York Times,* 4/10/80, pp. D1; D12.

Island Creek Coal Company
Sources

Burch, Elza, Head of Coal Preparation Plant Division, Island Creek Coal Company. Telephone interview with INFORM, 9/29/80.

Katlic, J.E., Vice President of Adminstration, Island Creek Coal Co., Correspondence to, and telephone conversation with, INFORM.

Moody's 1980 Industrial Manual "Occidental Petroleum Corporation," pp. 3773-3779.

Occidental Petroleum Corporation 1980 Annual Report and Form 10-K.

Monterey Coal Company
Sources

Davis, Harold, "Monterey No. 2 Builds Toward Its Tonnage Goal," *Coal Age Operating Handbook of Coal Preparation.* Edited by Paul C. Merritt (NY: McGraw Hill, 1978), pp. 160-162.

Exxon Corporation 1980 Annual Report and Form 10-K.

Mekelburg, Thomas A., Coal Preparation Engineer, Monterey Coal Company. Telephone interview with INFORM, 7/24/81.

Reid, F. Lynn, Employee Relations Manager, Monterey Coal Company. Telephone interview with INFORM, 7/24/81.

Schmidt, R., *Coal Preparation Plants* (Palo Alto, CA: Electric Power Research Institute, July 1978).

Northwest Coal Corporation
Sources

Amundsen, Al, Chief Engineer, Northwest Coal Corporation. Interview with INFORM, 8/14/81.

Jackson, Dan, "High Grade Western Coal Goes East," *Coal Age,* June 1980, pp. 62-68.

Matthies, Peter, Vice President, Northwest Energy Company. Interview with INFORM, 11/13/80.

Moody's 1980 Public Utility Manual, "Northwest Energy Company," pp. 2405-2408.

Northwest Energy Company 1980 Annual Report and Form 10-K.

Peabody Coal Company
Sources

Coal Age, "Peabody Adds Heavy Media Plant," October 1980, pp. 130-137.

Morris, George, Director, Preparation, Peabody Coal Company. Telephone interview with INFORM, 1/17/80.

Newmont Mining Corporation 1980 Annual Report and Form 10-K.

The Pittston Company
Sources

Business Week, "Pittston: Counting on 'Clean' Coal to Reverse the Tumble in Profits," September 8, 1980.

Coal Age, "Rum Creek Plant Blends Up to Six Seams," June 1980, pp. 15-16.

Denton, George H., Vice President, Technical Sales, The Pittston Company. Interview with INFORM, 10/1/80.

Harrold, Robert, "The Grand Badger Plant: Pittston's Design for the 1980's," *Coal Age,* December 1979, pp. 102-104.

Hensley, George H., Executive Vice President, The Pittston Company. Interview with INFORM, 10/1/80.

Jones, Don, Manager of Coal Preparation, The Pittston Company. Telephone interview with INFORM, 9/24/81.

The Pittston Company 1980 Annual Report and Form 10-K.

Salpukas, Agis, "Coal Boom Clogging Port," *New York Times,* 4/17/81, pp. D1; D6.

R&F Coal Company
Sources

Jung, Arthur D., "Determination of Attenuation of Sulfur Variability by Coal Preparation" (Unpublished draft. Springfield, VA: Versar, Inc.), pp. 4.1-4.20.

Moody's 1980 Industrial Manual, "Shell Oil Company," pp. 4004-4009.

Shell Oil Company 1980 Annual Report and Form 10-K.

Shores, Ken, Quality Control Superintendent and Lab Control Supervisor, R&F. Written Correspondence to INFORM, 10/24/80.

Spiker, William E., Vice President-Administration, R&F. Telephone interview with INFORM, 1/20/81.

American Electric Power (AEP)
Sources

American Electric Power Company 1980 Annual Report and Form 10-K.

American Electric Power Fact Sheets:
  Subject: Financial Performance 3/81
  Subject: Power Plants 3/81
  Subject: Coal 5/81
  Subject: Energy Mix, 1980

Big Business Day, "Utility: American Electric Power," 4/30/80 pp. 58-61.

Blackmore, Gerald, Executive Vice President, AEP, *The Outlook for Steam Coal As Used By Electric Utilities* (Paper presented to the American Mining Congress 1980 Mining Convention, San Francisco, September 21-24, 1980).

Feeny, Andy, "Taming the Giant," *The Power Line* (Washington, DC: Environmental Action Foundation) January 1980, pp. 6-8.

Merritt, Paul, C., ed., "AEP Washes Coal in Utah," *Coal Age Operating Handbook of Coal Preparation* (Coal Age Library of Operating Handbooks Vol. 3. N.Y: McGraw Hill, 1978), pp. 182-183.

Moody's 1980 Public Utility Manual, "American Electric Power Company, Inc.," pp. 933-940.

Duquesne Light Company
Sources

Duquesne Light Company 1980 Annual Report and Form 10-K.

Goodridge, Edward R., "Duquesne Light Maximizes Coal Recovery at Warwick Plant," *Coal Age Operating Handbook of Coal Preparation.* Edited by Paul C. Merritt (Coal Age Library of Operating Handbooks Vol. 3. NY: McGraw Hill, 1978), pp. 131-135.

INFORM interview with spokesperson from Duquesne Light, 11/13/80.

Johnson, Leroy W., Superintendent of Technical Services, Power Station Department, Duquesne Light Company. Response to the INFORM questionnaire.

Moody's 1980 Public Utility Manual, "Duquesne Light Company," pp. 1213-1218.

Pennsylvania Electric Company (PENELEC)
Sources

Esposito, Nicholas T. and Gray, Kenneth M., "Coal Cleaning Processes, Trials and Tribulations Between Conceptual Design and Full Scale Operation of a Prototype Design" (Homer City, PA: Pennsylvania Electric Company).

Merritt, Paul C., ed., "Multi-Stream Coal Cleaning System Promises Help with Sulfur Problem," *Coal Age Operating Handbook of Coal Preparation* (NY: McGraw Hill, 1978), pp. 59-61.

Meyer, Alden, "GPU On The Ropes," *The Power Line* (Washington, DC: Environmental Action Foundation, November 1980, Vol. 6, No. 4), pp. 1, 6-7.

Moody's 1980 Industrial Manual, "Pennsylvania Electric Company," pp. 1288-1291.

Pennsylvania Electric Company 1980 Annual Report and Form 10-K.

Pennsylvania Electric Company Fact Sheets:
Homer City Coal Cleaning Plant.
Some Questions and Answers About the Multi-Stream Coal Cleaning System.
Coal . . . Meeting Consumer Needs.

Pruce, Les, "Coal Cleaning at Homer City: An Alternative to Scrubbers," *Power* (NY: McGraw Hill, November 1978), pp. 213-216.

Tice, James, Manager of PENELEC's EPA/Electric Power Research Institute, Research and Development. Interview with INFORM, 9/18/80.

Pennsylvania Power and Light Company (PP&L)
Sources

Burkhart, V. R., Manager of Mining Technology and Engineering, PP&L. Interview with INFORM, 9/29/80.

Davis, Harold, "PP&L Updates its Preparation Plant to Reduce Coal's Sulfur Content by 40%" *Coal Age Operating Handbook of Coal Preparation.* Edited by Paul C. Merritt (NY: McGraw Hill, 1978), pp. 152-156; 210-211.

Herron, Richard, President, Pennsylvania Mines Corporation. Interview with INFORM, 9/19/80.

Moody's 1980 Public Utility Manual, "Pennsylvania Power and Light Company," pp. 3288-3290.

Pennsylvania Mines Corporation 1979 Annual Report and Form 10-K.

Pennsylvania Power and Light Company 1980 Annual Report and Form 10-K.

Tennessee Valley Authority (TVA)
Sources

Coal Age, "Another Snag for TVA Small Operation Plan: Unwashable Coal," July 1980, p. 33.

Cole, Randy, Project Manager, Combustion Systems Coal-Cleaning, TVA. Interview with INFORM, 9/8/80.

Frank, Robert L., Project Manager of Coal Cleaning and Reliability, TVA. Interview with INFORM, 9/8/80.

Hollinden, G. A., Project Manager of Environmental Control Technology, TVA. Interview with INFORM, 9/8/80.

Hollinden, G. A. and Massey, C. L., "TVA Compliance Programs for SO_2 Emissions" (Paper mimeographed by TVA, 1979).

Moody's 1980 Public Utility Manual, "Tennessee Valley Authority," pp. 3459-3467; a64, a66, a67.

Moore, Neil, Chief, Financial Planning and Management Staff, TVA. Interview with INFORM, 9/8/80.

Tennessee Valley Authority 1980 Annual Power Report.

Tennessee Valley Authority, "Paradise Stream Plant" (Technical Drawings and Flowsheet of TVA's Paradise Coal Washing Facility, 10/4/80).

Associated Electric Cooperative, Inc.
Sources

Associated Electric Cooperative, Inc., *Thomas Hill Energy Center* (pamphlet).

Associated Electric Cooperative, Inc., "New Madrid EPA Suit Dismissed," *Power News* (newsletter published by Associated Electric Cooperative, Inc.), November 1980, pp. 1, 3, 5.

Associated Electric Cooperative 1980 Annual Report and Form 10-K.

Brummett, Richard, Director of Fuel and Mining Division, Associated Electric Cooperative, Inc. Interview with INFORM, 11/12/80.

Needham, Rita, Coal Buyer, Associated Electric Cooperative, Inc. Interview with INFORM, 11/12/80.

Carolina Power and Light Company (CP&L)
Sources

Carolina Power and Light 1980 Annual Report and Form 10-K.

Carolina Power and Light, "Guide for CP&L Coal Suppliers" (pamphlet).

Moody's 1980 Public Utility Manual, "Carolina Power and Light Company," pp. 1061-1066.

Yarger, L.L., Manager, Fossil Fuels, CP&L. Interview with INFORM, 9/15/80.

Central Illinois Light Company (CILCO)
Sources

Central Illinois Light Company 1980 Annual Report and Form 10-K.

Grigsby, Leo F., Fuel and Contracts Manager, CILCO. Interview with INFORM, 11/6/80.

Herren, Robert, Manager of General Services, CILCO. Interview with INFORM, 11/6/80.

Moody's 1980 Public Utility Manual, "Central Illinois Light Company," pp. 1074-1078.

New England Electric System (NEES)
Sources

Knight, Michael, "Utility Contracts to Buy Ship to Transport Coal," *New York Times,* 11/1/80.

Lawrence, Damon, Manager of Coal Supply, NEES. Interview with INFORM, 1/30/80.

Moody's 1980 Public Utility Manual, "New England Electric System," pp. 3163-3188.

New England Electric System, pub., *NEESPLAN.*

New England Electric System, pub., *Coal Conversion at Brayton Point.*

New England Electric System 1980 Annual Report, Statistical Report, and Form 10-K.

Shenon, Philip W., "Many Electric Utilities Suffer as Conservation Holds Down Demand," *Wall Street Journal,* 10/9/80.

Work Group on Conversion to Coal at Brayton Point, *Conversion To Coal at the Brayton Point Power Plant in Somerset, Massachusetts* (Final report presented to the New England Energy Task Force, October 1978).

Public Service Company of Indiana
Sources

Aimone, Gene, Fuel Procurement Supervisor, Public Service Company of Indiana. Interview with INFORM, 9/25/80.

Moody's 1980 Public Utility Manual, "Public Service Company of Indiana, Inc.," pp. 2597-2605.

Public Service Indiana 1980 Annual Report and Form 10-K.

Iowa State University-Ames Laboratory
Sources

Bechtel Corporation, *Environmental Control Implications of Generating Electric Power from Coal: 1977 Technology Status Report, Appendix A (Part 1): Coal Preparation and Cleaning Assessment Study* (Prepared for Argonne National Laboratory; Work sponsored by the Division of Environmental Control Technology, U.S. Department of Energy, ANL/ECT-3, December 1977), pp. 260-283.

Capes, C.E., Smith, A.E. and Puddington, I.E., "Economic Assessment of the Application of Oil Agglomeration to Coal Preparation," *CIM Bulletin* (Montreal, Canada: Canadian Institute of Mining and Metallurgy, July 1974), pp. 115-119.

Leonard, W.M., Greer, R.T., Markuszewski, R. and Wheelock, T.D., "Coal Desulfurization and Deashing by Oil Agglomeration" (Paper presented at the Second Symposium on Separation Science and Technology for Energy Applications, Gatlinburg, Tennessee, May 5-8, 1981).

Patterson, E.C., Le, H.V., Ho, T.K., and Wheelock, T.D., "Better Separation by Froth Flotation and Oil Agglomeration," *Coal Processing Technology, Volume V* (NY: American Society of Chemical Engineers, 1979), pp. 171-177.

Wheelock, Thomas D., Professor, Iowa State University. Response to the INFORM questionnaire.

Wheelock, T.D., and Ho, T.K., "Modification of the Flotability of Coal Pyrites" (Paper presented at the Society of Mining Engineers of AIME, Annual Meeting, New Orleans, LA, February 18-22, 1979).

Wheelock, T.D. and Markuszewski, R., "Fossil Energy Annual Report: October 1, 1978-September 30, 1979; Section A: Advanced Development of Fine Coal Desulfurization and Recovery Technology " (Ames, IA: Ames Laboratory, Iowa State University).

Wheelock, T.D. and Markuszewski, R., "Physical and Chemical Coal Cleaning " (Paper presented at the Conference on the Chemistry and Physics of Coal Utilization, Energy Research Center, West Virginia University, Morgantown, WV, June 2-4, 1980).

Helix Technology Corporation
Sources

Ban, Thomas E., "Converting High-Sulfur Coal to Clean Pellet Fuel," *Mining Congress Journal,* May 1980, pp. 25-28.

Ban, Thomas E., "Converting High-Sulfur Coal to Clean Pellet Fuel," *American Mining Congress Convention News* (Paper presented at the 1979 American Mining Congress Convention, Los Angeles, CA, September 23-26).

Ban, Thomas E., Johnson, Eric and Marlowe, William E., "Carbonization Process for the Production of Clean Pellet Fuel," *Skillings Mining Review,* Vol. 69 No. 3, January 19, 1980.

Ban, T.E. Johnson, E.K. and Rodgers, L.W., "Low-Btu Gas from an Environmentally Clean Pellet Fuel," *American Society of Mechanical Engineers* (Paper presented at Winter Annual Meeting, Chicago, Illinois, November 16-21, 1980).

Brody, Christopher, Research Engineer, McDowell-Wellman. Interview with INFORM, 11/14/80.

Oak Ridge National Laboratory
Sources

Coal Age, "High Gradient, Open Gradient Systems Have Marked Potential," January 1980, pp. 73-79.

Coal Age, "Oak Ridge Lab Says Dry Coal Preparation Better Than Wet," June 1980, p. 29.

Hise, E.C., Group Leader, Coal Preparation and Waste Disposal, Engineering Technology Division, Oak Ridge National Laboratory. Telephone interview with INFORM, 12/5/80.

Hise, E.C., "Correlation of Physical Coal Separations—Part I" (Oak Ridge National Laboratory Publication No. ORNL-5570, September 1979).

Hise, E.C., "Sulfur and Ash Reduction in Coal by Magnetic Separation" (Paper presented at the Second U.S. Department of Energy Environmental Control Symposium, March 17-19, 1980, Sheraton International Conference Center, Reston, VA).

Hise, E.C., "Coal Preparation and Waste Utilization" (Oak Ridge National Laboratory Monthly Reports, March 31, 1980 and June 30, 1980).

Hise, E.C., Wechsler, I. and Doulin, J.M., "Separation of Dry Crushed Coals by High-Gradient Magnetic Separation" (Oak Ridge National Laboratory Publication No. ORNL-5571, October 1979).

Karlson, Frederick V., Clifford, Kenneth L., Slaughter, William W. and Huettenhain, Horst, "Potential of Magnetic Separation in Coal Cleaning," *Proceedings: Symposium on Coal Cleaning to Achieve Energy and Environmental Goals—September 1978, Hollywood, FL—Volume I* (US Environmental Protection Agency: EPA-600/7-79-098a), pp. 568-598.

Krause, Carolyn, "Coal Cleaning," *Oak Ridge National Laboratory Review,* Winter 1980, pp. 18-23.

Otisca Industries, Ltd.
Sources

Coal Age, "Otisca Process Enters Demonstration Plant Phase," January 1980, pp. 79-83.

Keller, D.V. Jr., Smith, Clay D. and Burch, Elza F., "Demonstration Plant Test Results of the Otisca Process Heavy Liquid Beneficiation of Coal" (Paper presented to the Annual SME-AIME Conference, Atlanta, Georgia, March 7, 1977).

Otisca Industries, Ltd. Response to the INFORM questionnaire.

Smith, Clay D., "Otisca Process Goes On-Line" (Paper presented to the Coal Age Conference, Louisville, Kentucky, October 23, 1979).

Other Technologies
Sources

Advanced Energy Dynamics, "Business Plan—Chapters 1, 2 and 3" (In-house position paper, 1981).

Bechtel Corporation, *Environmental Control Implications of Generating Electric Power from Coal: 1977 Technology Status Report, Appendix A, Part I, Coal Cleaning Preparation and Cleaning Assessment Study* (Prepared for Argonne National Laboratory; Work sponsored by the Division of Environmental Control Technology, U.S. Department of Energy, ANL/ECT-3, December 1977), pp. 206-224.

Coal Age, "Australian Process Super Cleans Fine Coal," January 1980, pp. 83-84.

Funston, G. Keith, Jr., Vice President of Finance and Administration, Advanced Energy Dynamics, Inc., Natick, Massachusetts. Telephone interview with INFORM, 9/11/81.

Lloyd, Alan J., "Clean Coal—New Facts," *ERT,* September 1980, pp. 9-17.

Stone and Webster Engineering Corporation, "Economic Analysis Advanced Energy Dynamics Coal Cleaning System" (Prepared for Advanced Energy Dynamics, June 18, 1980).

Electric Power Research Institute (EPRI)
Sources

Carlson, Fred, Program Manager, Coal Quality, EPRI. Interview with INFORM, 10/28/81.

Clifford, Kenneth, former Project Manager, Coal Quality, EPRI. Interview with INFORM.

Degner, V.R. and Olson, T.J., *Quarterly Report No. 2—Fine Coal Preparation Advancement Project (Flotation)* (WEMCO Division, Envirotech Corporation, December 1979).

EPRI in-house list of Coal Cleaning R&D Projects.

EPRI Journal, "More Coal Per Ton," June 1979, pp. 6-13.

EPRI Journal, "Coal Cleaning," November 1979, pp. 43-43.

EPRI Journal, "Coal Cleaning Facility Gears Up," September 1981, p. 30.

Rogers, Reed S.C., "Size Reduction of Coal and the Liberation of Pyrite," *Kennedy Van Saun Corporation—Quarterly Progress Report for the Period November-December, 1979.*

Sehgal, Randhir, Project Manager, Coal Quality, EPRI, Interview with INFORM, 11/25/80.

Van Atta, Daniel, Public Relations, EPRI. Interview with INFORM, 10/28/81.

Yeager, Kurt, "R&D Status Report, Coal Combustion Systems Division," *EPRI Journal,* June 1980, pp. 51-52.

U.S. Department of Energy (DOE)
Sources

Burger, John R., "DOE Launches Coal Preparation Research," *Coal Age,* June 1980, p. 102-107.

Crane, David, former member of the Office of Project Management, DOE. Interview with INFORM, 10/10/80.

Dietz, Albert G., Jr., Program Manager, Coal Preparation, DOE. Interview with INFORM, 10/7/80.

Frye, Keith, Acting Deputy Director, Office of Coal Processing, DOE. Interview with INFORM, 10/10/80.

U.S. Department of Energy, "Coal Preparation" (DOE draft paper, 1980), pp. 107-130.

U.S. Environmental Protection Agency (EPA)
Sources

Kilgroe, James, Coal Cleaning Program Manager, Office of Coal Processing, EPA. Interview with INFORM, 9/10/80.

U.S. Environmental Protection Agency, *First Progress Report: Physical Coal-Cleaning Demonstration at Homer City, Pennsylvania* (U.S. Environmental Protection Agency: EPA-625/2-79-023, August 1979).

U.S. Environmental Protection Agency, "Coal Cleaning and Related Areas," *Industrial Environmental Research Laboratory—Research Triangle Park Reports* (U.S. Environmental Protection Agency, July 31, 1980).

Appendix
Preparation Plant Wastes
Notes

1. National Coal Association, *Coal Data 1978* (Washington, DC: National Coal Association, 1980), pp. II-39 to II-40.

2. Office of Technology Assessment, *The Direct Use of Coal* (Washington, DC: Office of Technology Assessment, April 1979), p. 251.

3. W.A. Wahler and Associates, *Pollution Control Guidelines for Coal Refuse Piles and Slurry Ponds* (U.S. Environmental Protection Agency, EPA-600/7-78-222, November 1978), p. 38.

4. Office of Technology Assessment, *The Direct Use of Coal,* p. 251.

5. Appalachian Regional Commission, "Acid Mine Drainage in Appalachian," (Washington, DC: Appalachian Regional Commission, 1969), p. 6.

6. Ibid.

7. Wahler, *Pollution Control Guidelines,* p. 16.

8. E.M. Wewerka et al., "Control of Trace Element Leaching from Coal Preparation Wastes" *U.S. EPA Proceedings: Symposium on Coal Cleaning to Achieve Energy and Environmental Goals—September, 1978, Hollywood, FL, Vol. II* (U.S. Environmental Protection Agency, EPA-600/7-79-089b, April 1979), p. 861.

9. Ibid.

10. K.B. Randolph et. al., "Characterization of Preparation Plant Wastewaters," *U.S. EPA Proceedings: Symposium on Coal Cleaning,* pp. 828-829.

11. Wewerka, "Control of Trace Element Leaching," p. 861.

12. Office of Technology Assessment, *The Direct Use of Coal,* p. 234.

13. Ibid.

14. Personal communication with James Kilgroe, U.S. Environmental Protection Agency, 9/23/81.

15. R.M. Schuller et. al., "Chemical and Biological Characterization of Leachate from Coal Cleaning Wastes," *U.S. EPA Proceedings: Symposium on Coal Cleaning,* p. 919.

16. Wahler, *Pollution Control Guidelines,* p. 17.

17. Albert W. Duerbrouck and Richard Hucko, "Coal Cleaning and Desulfurization," *The Direct Use of Coal, Volume II—Part A: Working Papers, Appendix IV* (Washington, DC: Office of Technology Assessment, April 1979), p. 24.

18. Wahler, *Pollution Control Guidelines,* p. 17.

19. Office of Technology Assessment, *The Direct Use of Coal,* p. 235.

20. Wahler, *Pollution Control Guidelines,* pp. 18-19.

21. Richard Brown, ed., *Health and Environmental Effects of Coal Technologies: Background Information on Processes and Pollutants* (The MITRE Corporation, Metrek Division, DOE/HEW/EPA-04; MTR-79W0015901), p. 46.

22.. Wahler, *Pollution Control Guidelines,* pp. 18-19.

23. Ibid., pp. 21-43.

24. David R. Maneval, "Recent European Practice In Coal Refuse Utilization" (Paper delivered at the First Kentucky Coal Refuse Disposal and Utilization Seminar, Cumberland, Kentucky, May 2, 1975), p. 2.

25.. Wahler, *Pollution Control Guidelines,* pp. 40-43.

26. Personal communication with Randhir Sehgal, Electric Power Research Institute, 9/22/81.

27. Wahler, *Pollution Control Guidelines,* pp. 22-43.

28. Personal communication with Randhir Sehgal, 9/22/81.

29. Maneval, "Recent European Practice in Coal Refuse Utilization," p. 8.

30. Wahler, *Pollution Control Guidelines,* pp. 28-31, 40-43.

31.. Personal communication with George Sall, U.S. Department of Energy, 9/23/81.

32. Wahler, *Pollution Control Guidelines,* pp. 11-14.

33. Alan T. Law and Eugene Kitts, "Refuse Disposal Costs—Complying With Current Regulations" (Paper presented at the 1981 American Mining Coal Convention, Coal Preparation—Environmental Controls Session, May 12, 1981), pp. 2-3.

34. Ibid., p. 3.

35. Wahler, *Pollution Control Guidelines,* pp. 61-63, 74-76.

36.. Bechtel Corporation, *Environmental Control Implications of Generating Electric Power from Coal, 1977 Technology Status Report, Appendix A (Part I): Coal Preparation and Cleaning Assessment Study* (Prepared for Argonne National Laboratory; Work sponsored by the Division of Environmental Control Technology, U.S. Department of Energy, ANL/ECT-3, December 1977), p. 506.

37. Wahler, *Pollution Control Guidelines*, pp. 61-63, 74-76.

38. Ibid., p. 80.

39. Ibid.

40. Ibid., pp. 80-81.

41. Gibbs and Hill, Inc., *Coal Preparation for Combustion and Conversion* (Palo Alto, CA: Electric Power Research Institute, EPRI AF-791, Project 466-1, Final Report, May 1979), p. 7-4.

42. Personal communication with James Kilgroe, U.S. Environmental Protection Agency, 10/1/81.

43. Wahler, *Pollution Control Guidelines*, p. 99.

44. Personal communication with James Kilgroe, 10/1/81.

45. Wahler, *Pollution Control Guidelines*, p. 85-92.

46. Bechtel Corporation, *Environmental Control Implications of Generating Electric Power form Coal*, pp. 506-512.

47. Ibid., pp. 512-515.

48. Wahler, *Pollution Control Guidelines*, p. 99.

49. Ibid.

50. Bechtel Corporation, *Environmental Control Implications of Generating Electric Power from Coal*, pp. 512-513.

51. Wahler, *Pollution Control Guidelines*, p. 99.

52. P. Wagner et. al., "Trace Element Contamination of Drainage from Coal Cleaning Wastes," *Coal Processing Technology*, Vol. IV, 1980, p. 82.

53. Lawrence E. Wangen et al., "Control Technology for Coal Cleaning Wastes" (Reprinted from the Workshop on Solid Waste Research and Development Needs for Emerging Coal Technologies, American Society of Chemical Engineers, San Diego, CA, April 23-25 1979), p. 86.

54. Wagner, "Trace Element Contamination," pp. 82-83.

55. P. Wagner, *Environmental Control Technology Reserach In High Sulfur Coal Preparation Waste Drainages* (Paper prepared under contract for the U.S. Energy Research and Development Administration by the University of California Los Alamos Scientific Laboratory, 1980), pp. 3-7.

56. Edward F. Thode et al., *Costs of Coal and Electric Power Production—The Impact of Environmental Control Technologies for Coal-Cleaning Plants* (Los Alamos, NM: University of California Los Alamos Scientific Laboratory, 1979), pp. 8-9.

57. Wangen, "Control Technology for Coal Cleaning Wastes," p. 86.

58. Wagner, "Trace Element Contamination," p. 83.

59. Wagner, "Environmental Control Technology Research," pp. 6-7.

60. Thode, "Costs of Coal and Electric Power Production," pp. 8-9.

61. Wagner, "Environmental Control Technology Research," pp. 7-10.

62. Wagner, "Trace Element Contamination," p. 83-85.

63. Wangen, "Control Technology for Coal Cleaning Wastes," pp. 8-9.

64. Thode, "Costs of Coal and Electric Power Production," pp. 8-9.

65. Wangen, "Control Technology for Coal Cleaning Wastes," pp. 83.

66. Wagner, "Trace Element Contamination," p. 84.

67. Wangen, "Control Technology for Coal Cleaning Wastes," pp. 84-86.

68. Wagner, "Environmental Control Technology Research," p. 14.

69. Wagner, "Trace Element Contamination," p. 85.

70. Wangen, "Control Technology for Coal Cleaning Wastes," pp. 87-88.

71. Wagner, "Trace Element Contamination," p. 85.

72. Ibid.

73. Wangen, "Control Technology for Coal Cleaning Wastes," pp. 88

74. Thode, "Costs of Coal and Electric Power Production," pp. 8-9.

75. Wagner, "Environmental Control Technology Research," p. 15.

76. Wangen, "Control Technology for Coal Cleaning Wastes," p. 90

77. David R. Maneval, 'Recent Developments In Reprocessing Of Refuse For A Second Yield of Coal'' (Paper delivered at the Third Symposium on Coal Preparation, NCA/BCR Coal Conference and Expo IV, Louisville, Kentucky, October 18-20, 1977), p. 156.

78. Richard H. Mason, "Coal By-Products Constiute Highway Base," *Coal Mining & Processing*, May 1974, p. 45.

79. Bechtel Corporation, *Environmental Control Implications of Generating Electric Power from Coal*, p. 512.

80. Jerry G. Rose, "New Uses for Coal Refuse as Construction Materials," *Mining Congress Journal*, Vol. 66, No. 9, May 1974, p. 45.

81. Maneval, "Recent Developments in Reprocessing of Refuse," pp. 152-157.

82. Ibid.

83. *Business Week,* "Coal Wastes' Many Treasures," January 14, 1980, p. 118.

84. Ralph E. Brown, T.C. Wilson and David L. Thomasson, "Economic Evaluation of Coal Refuse Disposal Systems" ((Paper presented at the Third Symposium on Coal Preparation, NCA/BCR Coal Conference and Expo IV, Louisville, Kentucky, October 18-20, 1977), pp. 141-151.

85. Ibid.

86. Bechtel Corporation, *Environmental Control Implications of Generating Electric Power from Coal,* p. 512.

87. Office of Technology Assessment, *The Direct Use of Coal,* p. 251.

88. John K. Alderman and William M. Smith, "Acid Mine Drainage: the Problem and the Solution," *Coal Mining & Processing,* August 1977, pp. 66-68, 87-88.

89. Jean Moore, D. Sargent and A. Sapsaloupoulou, "Coal Cleaning Facilities: A Summary of Environmental Regulations" (Revised draft prepared for the U.S. Environmental Protection Agency, Office of Research and Development, Washington, DC, March 24, 1981), pp. 5-6.

90. Ibid., pp. 6, 8.

91. Ibid., pp. 12-17.

92. Ibid., pp. 17-19.

93. Ibid., pp. 8-12.

94. Ibid., pp. 19-20.

95. Ibid., p. 7.

Appendix
Amenability of Coal to Preparation
Notes

1. Gibbs and Hill, Inc., *Coal Preparation for Combustion and Conversion* (Palo Alto, CA: Electric Power Research Institute, EPRI AF-791, Project 466-1, Final Report, May 1978), p. 2-17.

2. Howard W. Decker, Jr. and John N. Hoffman, *Coal Preparation Volume I* (University Park, PA: The Pennsylvania State University, 1963). p. 125.

3. Gibbs and Hill, Inc., *Coal Preparation for Combustion and Conversion,* p. 2-28.

4. U.S. Department of the Interior, Bureau of Mines, *Sulfur Reduction Potential of the Coals of the United States* (Washington, DC: U.S. Bureau of Mines, Report of Investigations/ 1976 RI:8118), pp. 4-6.

5. Gibbs and Hill, Inc., *Coal Preparation for Combustion and Conversion,* p. 2-17.

6. Decker and Hoffman, *Coal Preparation Volume I,* p. 131.

7. Ibid., p. 131.

8. Gibbs and Hill, Inc., *Coal Preparation for Combustion and Conversion,* p. 2-19.

9. Ibid., pp. 2-19 to 2-21.

10. Decker and Hoffman, *Coal Preparation Volume I,* p. 138.

11. The President's Commission on Coal, *Coal Data Book* (Washington, DC: U.S. Government Printing Office, February 1980), p. 71.

12. Gibbs and Hill, Inc., *Coal Preparation for Combustion and Conversion,* pp. 2-42, 2-44.

13. Ibid.

14. U.S. Department of the Interior, Bureau of Mines, *Sulfur Reduction Potential of the Coals of the United States,* p. 1.

15. James D. Kilgroe, "Combined Coal Cleaning and FGD," *Symposium on Flue Gas Desulfurization—Las Vegas, Nevada, March 1979; Volume I* (U.S. Environmental Protection Agency: EPA-600/7-79-167a, July 1979), pp. 9, 12, 16.

16. Ibid.

17. U.S. Department of the Interior, Bureau of Mines, *Sulfur Reduction Potential of the Coals of the United States,* pp. 29-30.

18. Ibid., p. 30.

19. Marcella A. Wells et. al., *Coal Resources and Sulfur Emission Regulations: A Background Document* (Berkeley, CA: Teknekron Research, Inc., Report No. R-031-VER-79/R-3, February 1980).

20. Ibid.

Appendix
Description of Coal
Notes

1. The President's Commission on Coal, *Coal Data Book* (Washington, DC: U.S. Government Printing Office, Februrary 1980), p. 60.

2. Office of Technology Assessment, *The Direct Use of Caol* (Washington, DC: Office of Technology Assessment, April 1979), p. 60.

3. Gibbs and Hill, Inc., *Coal Preparation for Combustion and Conversion* (Palo Alto, CA: Electric Power Research Institute, EPRI AF-791, Project 466-1, Final Report, May 1978),p. 1-2.

4. The President's Commission on Coal, *Coal Data Book,* p. 71.

5. Howard W. Decker, Jr. and John N. Hoffman, *Coal Preparation Volume 1* (University Park, PA: The Pennsylvania State University, 1963), pp. 57-75.

6. Ibid.

7. Ibid.

8. Ibid.

9. Gibbs and Hill, Inc., *Coal Preparation for Combustion and Conversion,* p. 1-1.

10. Ibid., p. 1-4.

11. Ibid., p. 1-3.

12. Decker and Hoffman, *Coal Preparation,* p. 66.

13. Gibbs and Hill, Inc., *Coal Preparation for Combustion and Conversion,* pp. 1-6, 1-9.

14. Ibid., p. 1-8.

15. Decker and Hoffman, *Coal Preparation,* p. 80.

16. Gibbs and Hill, Inc., *Coal Preparation for Combustion and Conversion,* p. 1-8.

17. Ibid., p. 6-4.

18. Decker and Hoffman, *Coal Preparation,* p. 80.

19. Gibbs and Hill, Inc., *Coal Preparation for Combustion and Conversion,* p. 6-4.

20. Decker and Hoffman, *Coal Preparation,* p. 80.

21. K.L. Maloney, P.K. Engel and S.S. Cherry, *Sulfur Retention in Coal Ash* (U.S. Environmental Protection Agency; EPA-600/78-153b, November 1978), p. 1.

22. U.S. Department of the Interior, Bureau of Mines, *Sulfur Reduction Potential of the Coals of the United States* (Washington, DC: U.S. Bureau of Mines, Report of Investigations/1976 RI: 8118).

23. Ibid.

24. Albert W. Deurbrouck and Richard Hucko, "Coal Cleaning and Desulfurization," *The Direct Use of Coal, Volume II—Part A: Working Papers, Appendix IV* (Washington, DC: Office of Technology Assessment, April 1979), p. 25.

25. Harold L. Lovell, "Particle Size Distribution In the Liberation of Pyrite in Coal," *U.S. EPA Proceedings: Symposium on Coal Cleaning to Achieve Energy and Environmental Goals—September, 1978—Hollywood, FL—Vol. I* (U.S. Environmental Protection Agency: EPA—600/7-79-089a, April 1979), p. 93.

26. U.S. Department of the Interior, Bureau of Mines, *Sulfur Production Potential of the Coals of the United States,* p. 2.

27. Deurbrouck and Hucko, *Coal Cleaning and Desulfurization,* p. 25.

28. Gibbs and Hill, Inc. *Coal Preparation for Combustion and Conversion,* pp. 1-11 to 1-13.

29. J.A. Cavallaro and A.W. Deurbrouck, *A Washability and Analytical Evaluation of Potential Pollution from Trace Elements in Coal* (U.S. Environmental Protection Agency: EPA-600/7-78-038, March 1978).

30. Gibbs and Hill, Inc. *Coal Preparation for Combustion and Conversion,* p. 1-7.

31. Decker and Hoffman, *Coal Preparation,* p. 80.

32. Gibbs and Hill, Inc. *Coal Preparation for Combustion and Conversion,* p. 1-4.

33. Decker and Hoffman, *Coal Preparation,* p. 80.

34. Ibid., pp. 77-78.

35. Ibid., p. 78.

36. Gibbs and Hill, Inc. *Coal Preparation for Combustion and Conversion,* pp. 1-9 to 1-10.

37. Decker and Hoffman, *Coal Preparation,* p. 86.

38. *Ibid.,* pp. 67, 86; 6-7.

39. *Ibid.,* p. 87.

40. Gibbs and Hill, Inc. *Coal Preparation for Combustion and Conversion,* pp. 2-9, 2-12, 2-14.

41. Written communication from Mr. George W. Sall, Acting Director, Division of Coal Production Technologies, U.S. Department of Energy, to INFORM, May 7, 1981.

42. Decker and Hoffman, *Coal Preparation,* pp. 81-82.

43. Ibid., p. 82.

44. Ibid., pp. 83-84.

Appendix
U.S. Coal Deposits
Notes

1. Office of Technology Assessment, *The Direct Use of Coal* (Washington, DC: Office of Technology Assessment, April 1979), p. 55.

2. Ibid., p. 56.

3. Ibid., p. 56.

4. 1977 Keystone Coal Industry Manual (NY: McGraw-Hill, Mining Informational Services, 1979), p. 777.

5.   Office of Technology Assessment, *The Directo Use of Coal,* p. 56.

6.   *1977 Keystone Coal Industry Manual,* p. 780.

7.   Office of Technology Assessment, *The Direct Use of Coal,* p. 56.

8.   The President's Commission on Coal, *Coal Data Book* (Washington, DC: U.S. Government Printing Office, February 1980), pp. 62-63.

9.   Ibid., pp. 64-65

Glossary

acid drainage: The runoff of acidic liquids from coal-production waste piles. Such runoff can contaminate ground and surface waters.

acid rain: A solution of acidic compounds formed when sulfur and nitrogen oxides react with water droplets and airborne particles.

anthracite coal: A hard, lustrous, high-ranked coal, containing a high percentage of fixed carbon and a low percentage of moisture.

A/E/C firm (architectural, engineering and construction firm): A firm which is hired by coal companies and utilities to design, engineer and construct coal cleaning facilities.

ash: The noncombustible residue remaining after coal has been completely burned, formed from the noncombustible minerals in coal.

ash fusion temperature: The temperature at which a coal ash fuses or melts. This is an important characteristic of coal, as the fused ash can interfere with the free flow of air necessary for proper combustion.

attainment area: Area that has met the National Ambient Air Quality Standards established by the Clean Air Act of 1970.

availability: Percentage of time that a system (or piece of equipment) is in working order, available for operation.

beneficiation: See *physical coal cleaning*.

Best Available Control Technology (BACT): A technology, determined on a case-by-case basis for major sources of pollution located in attainment areas (see *attainment area*), that is the most effective pollution control yet demonstrated.

baghouse: An air pollution control device that removes particulate matter from flue gas, usually achieving a removal rate above 99.9 percent.

barrel: Liquid volume measure equal to 42 gallons, commonly used in measuring petroleum or petroleum products.

bench scale: A small-scale laboratory unit for testing process concepts and operating parameters as the first step in the evaluation of a process.

bituminous coal: A soft coal, ranked below anthracite. It accounts for the bulk of all coal mined and cleaned in this country, and is the coal most often burned by utilities.

Btu: British thermal unit: the amount of energy needed to raise the temperature of one pound of water one degree Fahrenheit.

calorie: The amount of heat required to raise the temperature of 1 kilogram of water by 1° C at 1 atmosphere pressure.

captive mine: A mine owned by the user of the coal, most often an industry or utility.

chemical coal cleaning: An advanced cleaning process that breaks down the molecular structure of coal and removes most organic and inorganic sulfur.

Clean Air Act of 1970: As amended in 1977, a comprehensive federal statute that sets specific goals for the quality of the nation's air resources and authorizes the states to determine how these goals are to be met.

coal conversion: The process which enables a utility to switch from burning oil to burning coal by installing coal-handling, feeding, and boiler equipment and air pollution control equipment.

coal gasification: Production of synthetic gas from coal.

coal liquefaction: Conversion of coal to a liquid.

coal sizes: In the coal industry, the term "5 inch to ¾ inch" means all coal pieces between 5 inches and ¾ inch at their widest point. "Plus 5 inch" means coal pieces over 5 inches in size; "1½ to 0" or "-1½" means coal pieces 1½ inches and under.

coarse coal: Coal pieces larger than ½ millimeter in size.

cogeneration: A process by which electricity and steam, for space heating or industrial-process heating, are produced simultaneously from the same fuel.

coke: A porous, solid residue resulting from the incomplete combustion of coal, used primarily in the steel-making process.

corrosion: The gradual degrading and wearing away of a surface, as by rusting or by the action of chemicals.

cyclone: Equipment employed in the coal cleaning process that uses centrifugal force to separate coal from its impurities.

dedicated mine: A mine that has been developed by both the owner of the mine and the user of the coal, to supply that user only. (In the case of a utility that burns coal, the mine might supply a large generating station for its lifetime, usually 30 to 40 years.)

dewatering: The mechanical removal of moisture from coal slurries, used in all coal preparation plants employing wet washing technologies.

drying: The removal of water from coal by thermal drying, screening, or centrifuging.

electrostatic precipitator (ESP): An air pollution control device that uses an electric field to remove particulate matter from a gas stream.

fine coal: Coal pieces less than ½ millimeter in size.

flue gas desulfurization (FGD or scrubbing): The removal of sulfur oxides from stack gases of a coal-fired boiler.

fluidized-bed combustion (FBC): The process whereby coal is burned in a fluidized bed for greater efficiency and environmental control. Its basic principle involves the blowing of crushed coal for combustion into a bed of inert ash mixed with limestone or dolomite.

fly ash: Airborne bits of unburnable ash that are carried into the atmosphere by stack gases.

fouling: The accumulation of small, sticky molten particles of coal ash on a boiler surface.

free on board (FOB): Price of coal before the addition of transportation cost.

froth flotation: A coal cleaning process, using aerated water, that separates coal from its impurities by taking advantage of the difference between the surface properties of coal and its wastes.

gpm: Gallons per minute.

groundwater: Subsurface water found in aquifers (porous rock strata, soils, and underground rivers), from which wells and springs are fed.

heavy media: Liquids that have a greater density than water, which are used to separate coal from its impurities.

jig: A cleaning device that stratifies coal and refuse particles in water by means of pulsations, thus separating the coal from its impurities.

kilowatt: A unit of electrical power equal to 1,000 watts.

leach: To filter through some solid material.

leachate: Liquid formed by water seeping through potentially toxic solid materials.

level: An arbitrary term that helps categorize the degree to which coal is cleaned.

lignite: A soft, brownish-black, high-moisture coal that is the lowest ranked among coals.

magnetite: A finely ground iron oxide that, mixed with water, is often used as a separating medium in the heavy-media cleaning process.

megawatt: A unit of electrical power equal to 1,000,000 watts.

mesh: A measure of size often used to describe coal particles. The origin of the term comes from the number of open spaces in a screen or sieve per linear inch. "28 mesh" is equal to 595 millimeters, or .0234 inch. "-28 mesh" refers to all coal particles that are smaller than 28 mesh. "100 mesh" and "200 mesh" are extremely fine-size coal particles.

metallurgical coal: Coal used in the steel-making process to manufacture coke.

middlings: Coal of an intermediate specific gravity and quality.

NA: Not available.

New Source Performance Standards (NSPS): Emissions limitation standards established in 1971 by the administrator of the Environmental Protection Agency. NSPS apply to large new (or substantially modified) stationary sources of pollution that are fired by fossil fuel.

nitrogen dioxide (NO_2): A brown compound released as a pollutant in automobile exhaust and in the combustion of coal. It is a major contributor to photochemical smog and has been shown to aggrevate several respiratory diseases.

non-attainment area: Area that has not met the National Ambient Air Quality Standards established by the Clean Air Act of 1970.

pH: An indicator which expresses both the acidity and alkalinity of a given substance on a scale of 0 to 14. Seven represents neutrality; numbers below 7 indicate increasing acidity and numbers above 7 indicate increasing alkalinity.

peak load: The maximum demand for electrical energy (usually expressed in megawatts) for a given period of time.

peat: Partially carbonized plant matter, formed by slow decay in water.

physical coal cleaning: Processes which employ a number of different operations, including crushing, sizing, dewatering and clarifying, and drying, which improve the quality of the fuel by regulating its size and reducing the quantities of ash, sulfur, and other impurities. For purposes of this study, the term coal cleaning will be synonymous with the terms coal preparation, beneficiation, and washing.

plugging: The build-up of semi-soft solids in boiler tubes.

Prevention of Significant Deterioration (PSD): Program authorized by the Clean Air Act to prevent the degradation of air quality in clean-air areas. It specifies the maximum allowable increases in ambient concentrations of pollutants in these areas.

rank: A category for evaluating coal, ranging from lignite (soft) to anthracite (hard).

reagent: A substance used to bring about a chemical reaction.

recoverable reserves (coal): Unmined coal deposits that can be removed by current technology, taking into account economic, legal, political and social variables.

reserves (coal): All the coal deposits that are potentially mineable.

resources (coal): All the coal deposits present at a given location regardless of whether or not they can be mined.

Revised New Source Performance Standards: June 1979 revision of NSPS; stricter in some respects than NSPS.

run-of-mine (ROM) coal: Untreated coal as it leaves the mine.

scrubbers: See flue gas desulfurization.

seam: Underground layer of coal or other mineral of any thickness.

slagging: The accumulation of coal ash on the wall tubes of a coal-fired boiler furnace, forming a solid layer of ash residue and interfering with heat transfer.

slurry: A mixture of pulverized insoluble material and water.

slurry pipeline: A pipeline that can transport a coal-and-water mixture for long distances.

slurry pond: A waste-disposal site for fine-coal waste materials.

specification coal: Coal that has been prepared to meet user specifications that include size and sulfur, ash, Btu and moisture content.

specific gravity: The ratio of the density of a given substance to the density of water.

spot market: Commodity market where short-term purchases are made on a fixed-price basis with fixed terms and conditions.

sub-bituminous coal: A low-ranked coal, with high percentages of volatile matter and moisture, found mainly in the western United States.

sulfur: An element that is found in many fossil fuels.

sulfur dioxide (SO_2): An air pollutant usually generated by combustion of fuels that contain sulfur.

surface mining: Mining that exposes the coal seam by removing the soil and rocks above it.

synthetic fuel: A fuel produced by the biological, thermal or chemical transformation of other fuel or materials.

ten to the sixth power (10^6): 1,000,000; used in this study for emission standard formulas.

ten-year 24-hour precipitation event: A precipitation event or runoff from melting snow that is equal to the maximum amount of precipitation within a ten-year interval that would accumulate in 24 hours.

tph: Tons per hour.

trace element: Any element present in minute quantities, such as lead and mercury.

unit train: A railway train designed to achieve economies of scale by transporting a single commodity (such as coal), loading fully and operating nonstop.

volatile matter: In coal burning, the portions of the coal that are re-leased as gases — including carbon monoxide (CO), carbon dioxide (CO_2), and methane (CH_4) — when the coal is heated to a temperature of about 900^0 C.

washability analysis: A procedure used in a laboratory before prepa-ration plant design, to determine the cleaning processes to be employed, and used during normal operation, to evaluate the performance of the cleaning equipment and the amenability of the raw coal feed to the cleaning processes chosen.

watt: A unit of electrical activity or power — the power developed in a circuit by a current of one ampere flowing through a potential difference of one volt.

Bibliography

Bechtel Corporation. December 1976. *Environmental Control Implications of Generating Electric Power From Coal, Technology Status Report Volume II.* Prepared for Argonne National Laboratory; Work sponsored by the Division of Environmental Control Technology, US Department of Energy. ANL/ECT-1.

Bechtel Corporation. December 1977. *Environmental Control Implications of Generating Electric Power from Coal, 1977 Technology Status Report, Appendix A (Part 1) Coal Preparation and Cleaning Assessment Study.* Prepared for Argonne National Laboratory; Work sponsored by the Division of Environmental Control Technology, US Department of Energy. ANL/ECT-3.

Bechtel Corporation. June 1980. *An Analysis of Chemical Coal Cleaning Processes.* Prepared under contract for the US Department of Energy, Contract No. ET-78-C-01-3137.

Bechtel National, Inc. March 1981. *Impact of Coal Cleaning on the Cost of New Coal-Fired Power Generation.* Palo Alto, CA: Electric Power Research Institute. EPRI CS-1622, Project 1180-2, Final Report.

Brown, Richard, ed. August 1979. *Health and Environmental Effects of Coal Technologies*. McLean, VA: The MITRE Corporation, Metrek Division. DOE/HEW/EPA-04, MTR-79W0015901.

Buder, Manfred K. and Clifford, Kenneth L. 1979. "Cost Effects of Coal Cleaning on Utility Power Generation." Paper presented at Coal Technology '79–2nd International Coal Utilization Exhibition and Conference, Houston, TX, November 6-8, 1979.

Buder, M.K., Clifford, K.L., Huettenhain, H. and McGowin, C.R. 1979. "The Effects of Coal Cleaning on Power Generation Economics." Paper presented at the American Power Conference, Chicago, Illinois, April 23-25, 1979.

Contos, G.Y., Frankel, I.F., McCandless, L.C. August 1978. *Assessment of Coal Cleaning Technology: An Evaluation of Chemical Coal Cleaning and Low Sulfur Coal*. US Environmental Protection Agency: EPA-600/7-78-173a.

Decker, Howard W. and Hoffman, John N. 1963. *Coal Preparation, Volumes I and II*. University Park, PA: The Pennsylvania State University.

Duerbrouck, Albert W. and Hucko, Richard. March 1978. "Coal Cleaning and Desulfurization." Pittsburgh, PA: Department of Energy, Pittsburgh Mining Operation.

Gibbs and Hill, Inc. May 1978. *Coal Preparation for Combustion and Conversion*. Palo Alto, CA: Electric Power Research Institute. EPRI AF-791, May Project 466-1, Final Report.

Hoffman-Muntner Corporation. January 1978. *Engineering/Economic Analyses of Coal Preparation With SO₂ Cleanup Processes*. US Environmental Protection Agency: EPA-600/7-78-002.

Hoffman-Muntner Corporation. July 1978. *An Engineering/Economic Analysis of Coal Preparation Plant Operation and Cost*. US Environmental Protection Agency: EPA-600/7-78-124.

Hoffman-Muntner Corporation. Elmer C. Holt (author). 1980. *Effect of Coal Quality on Maintenance Costs At Utility Plants*. Report prepared for the US Department of Energy under Contract No. DEACO 1-75ET12512.

Keystone Coal Industry Manual. 1980. NY: McGraw Hill, Mining Informational Services.

Kilgroe, James D. July 1979. "Combined Coal Cleaning and FGD." *Symposium on Flue Gas Desulfurization–Las Vegas, Nevada, March 1979; Volume I*. US Environmental Protection Agency: EPA-600/7-79-167a.

Kilgroe, James D. and Strauss, Jerome B. 1980. "Uses of Coal Cleaning for Air Quality Management." Paper presented at the Second Conference on Air Quality Management in the Electric Power Industry, Center for Energy Studies, University of Texas at Austin, January 22-25, 1980.

McCandless, Lee C. and Shaver, Robert B. July 1978. *Assessment of Coal Cleaning Technology: First Annual Report*. US Environmental Protection Agency: EPA-600/7-78-150.

Merritt, Paul C., ed. 1978. *Coal Age Operating Handbook of Coal Preparation*. Coal Age Library of Operating Handbooks, Vol. 3. NY: McGraw Hill.

Moody's Industrial Manual. 1980. NY: Moody's Investors Service, Inc.

National Coal Association. 1980. *Coal Data 1978*. Washington, DC: National Coal Association.

National Coal Association. 1980. *International Coal 1979*. Washington, DC: National Coal Association.

National Coal Association. 1979. *Steam Electric Plant Factors 1979*. Washington, DC: National Coal Association.

Office of Technology Assessment. April 1979. *The Direct Use of Coal*. Washington, DC: Office of Technology Assessment.

PEDCo. Environmental, Inc. 1980. *Cost Benefits Associated With the Use of Physically Cleaned Coal*. US Environmental Protection Agency: Contract No. 68-02-2603; Task No. 31.

President's Commission on Coal. February 1980. *Coal Data Book*. Washington, DC: US Government Printing Office.

Schmidt, Richard A. July 1978. *Coal Preparation Plants*. Palo Alto, CA: Electric Power Research Institute.

Stone and Webster Engineering Corporation. April 1980. *Impact of Cleaned Coal on Power Plant Performance and Reliability*. Palo Alto, CA: Electric Power Research Institute. EPRI CS-1400, Project 1030-6, Final Report.

US Department of Energy, Energy Information Administration. 1981. *Monthly Energy Review: August 1981*. Washington, DC: US Government Printing Office. DOE/EIA-0035(81-08).

US Environmental Protection Agency. April 1979. *Proceedings: Symposium on Coal Cleaning to Achieve Energy and Environmental Goals—Hollywood, Florida, September 1978; Volumes I and II*. US Environmental Protection Agency: EPA-600/7-79-098a; 098b.

US Environmental Protection Agency and Tennessee Valley Authority. November 1979. *Evaluation of Physical/Chemical Coal Cleaning and Flue Gas Desulfurization*. US Environmental Protection Agency: EPA-600/7-79-250.

US Department of the Interior, Bureau of Mines. 1976. *Report of Investigations 1976, No. 8118, Sulfur Reduction Potential of the Coals of the United States*. Washington, DC: US Bureau of Mines.

W.A. Wahler and Associates. November 1978. *Pollution Control Guidelines for Coal Refuse Piles and Slurry Ponds*. US Environmental Protection Agency: EPA-600/7-78-222.

Wells, Marcella A. et. al. February 1980. *Coal Resources and Sulfur Emission Regulations: A Background Document*. Berkeley, CA: Teknekron Research Inc. Report No. R-031-VER-79/R-3.

Wilson, Carroll, Project Director, Massachusetts Institute of Technology. 1980. *Coal—Bridge to the Future: Report of the World Coal Study.* Cambridge, MA: Ballinger Publishing Company.

Methodololgy

Literature Review (November 1979 to 1980)

INFORM's project was initiated in November 1979 with a nine month review of literature on coal cleaning gathered from four primary areas:

Government documents from sources including the United States Environmental Protection Agency, the Department of Energy, the Department of the Interior–Bureau of Mines, state air and water pollution control agencies, the President's Commission on Coal, and the Congressional Office of Technology Assessment.

Industry materials from individual corporations and utilities operating coal preparation plants, from architect/engineering/ construction (A/E/C) firms that design and build them, and from relevant trade associations and research centers including the Electric Power Research Institute, the Edison Electric Institute, the National Coal Association and the Mining and Reclamation Congress.

Business and general publications from such industry periodicals as *Coal Age*, *Coal Week*, *Mining Congress Journal*, and *Engineering and Mining Journal*, as well as the *Wall Street Journal* and the *New York Times*. A list of key references used in this report is found in the Bibliography.

Selection of Project Advisory Board (June to July 1980)

A project advisory board was established which included representatives from business, government, public interest and independent research sectors. The role of this board was to offer advice on the directions of the project, the extent of research to be carried out and the final report manuscript (see list of advisors). Extensive consideration was given to the comments and advice of the board, although final responsibility for the content of this study resides with the authors and INFORM.

Questionnaires (August to September 1980)

Questionnaires for owners of coal preparation plants, architect/engineering/construction firms, utilities that burn cleaned coal, and research groups were developed defining the information INFORM would be seeking on the environmental and economic costs and benefits of coal cleaning. The questionnaires were reviewed by the advisory board before being distributed.

Selection of Sample Firms (August to September 1980)

INFORM selected 31 firms for specific study, including seven coal companies that own preparation plants, five utilities owning preparation plants and using cleaned coal, five utilities that buy cleaned coal from outside suppliers for use in their power plants, six A/E/C firms, six developers of specific advanced coal cleaning technologies not yet on the market, and three research groups conducting the most extensive research of advanced physical and chemical coal cleaning techniques. The preparation plant owners were chosen for the variety of types of coal they cleaned and the levels of cleaning provided. Only plants built after 1970 were selected for the study, since INFORM's primary concern was the definition of the state of the art and current practice. Five of the ten largest A/E/C firms, (as listed in the *Keystone Coal Industry Manual*) were selected, as well as Kaiser, a new entrant in this field. The ten utilities selected were either: 1) customers for the cleaned coal produced by the preparation plants in INFORM's sample (their views of the product could be compared to the views of the producers), or 2) utilities that cleaned coal for their own power generation needs. These utilities represent a varied sample in their geographical locations, the size and age of their boilers and their pollution control requirements.

Research Procedure (October 1980 to March 1981)

Most of the 31 firms in the sample responded to INFORM's questionnaire with interest and were willing to share their information and views on cleaned coal. Only three refused to cooperate: Peabody Coal, which in 1980 was the largest producer of coal in the United States, operating 28 coal cleaning plants; A.T. Massey, which operates 17 plants; and the American Electric Power Co., a utility that uses more cleaned coal than any other utility in the United States, and a utility that has conducted extensive research into the merits and drawbacks of cleaned coal. Information on these three firms' activities was, however, found in publicly available materials.

INFORM conducted interviews with technical experts and management representatives at each of the participating companies, and report authors toured the Homer City plant operated by PENELEC.

Report Preparation (March to August 1981)

Profiles of each firm's views and experience with cleaned coal were drafted, reviewed in house and then circulated to the respective companies to check for accuracy. Upon reviewing company comments, the profiles were revised and edited by INFORM's staff.

The information for the background chapters on coal cleaning technologies, plant waste disposal, the properties of coal, United States coal deposits, and the amenability of coal to cleaning, were obtained from the literature, from interviews with the companies in our sample and from reports and interviews with officials at DOE and EPA. These chapters were reviewed by the project advisors and revised by the authors.

Writing and Editing (July to December 1981)

The findings were based on information obtained from the interviews conducted by INFORM and from the current literature. Once again, these findings were reviewed by our project advisors and revised by INFORM's staff.

INFORM's staff accepts full responsibility for the final conclusions drawn from the research.

About the Authors

Cynthia A. Hutton
Project Director

Cynthia Hutton joined INFORM's energy staff in 1979 after researching and evaluating Congressional responses to the energy crisis at Har-
vard's J.F.K. School of Government. She was also a researcher at the
National Cancer Institute, Environmental Epidemiology Branch, investigating the implications of low-dose radiation. As a Congressional
intern, she researched and evaluated the possibilities of land use and
cogeneration in Milwaukee. She received her BA *cum laude* in Government and *summa cum laude* in Religion from Dartmouth College.

Robert N. Gould
Research Associate

Robert Gould joined INFORM's staff in June 1980 after receiving an
MPH degree from the Columbia University School of Public Health,
Division of Environmental Sciences, the first such degree to combine
social work and environmental sciences. His studies included molecular
and systemic toxicology, environmental law, research methodology, environmental health and energy/environmental planning and management. Mr. Gould holds a BA *magna cum laude* in Psychology from Con-
necticut College and an MS degree in Social Work from Columbia. He
has worked as a clinician and research assistant in several New York
and Massachusetts hospitals.

INFORM Publications

Newsletter

INFORM Reports: A bi-monthly newsletter reporting on INFORM's current research and educational activities. Articles and announcements.

$25 per annum (membership fee)

Air and Water Pollution

A CLEAR VIEW: Guide to Industrial Pollution Control (June 1975)
A manual on procedures for monitoring and assessing industrial air and water pollution problems and controls.

$6.95/$3.95 paperback

Nutrition

WHAT'S FOR DINNER TOMORROW? Corporate Views of New Food Products (January 1981)
An analysis of where, when, and to what degree nutritional considerations are factored into corporate decisions to develop and market a new food product.

$5.00

Energy

RECLAIMING THE WEST: The Coal Industry and Surface-Mined Lands (July 1980)
An analysis of the land reclamation practices of 13 coal companies at 15 surface mines in 6 western states. Defines best available reclamation techniques and evaluates the impact of each mine upon local land and water resources.
$45.00

ENERGY FUTURES: Industry and the New Technologies (July 1976)
An analysis of corporate research, commercial development, and environmental impact of new energy sources and technologies. (check availability before order)

INDUSTRIAL ENERGY CONSERVATION: Where Do We Go From Here? (December 1977)
A report on the federal programs and industry progress toward achieving federal energy conservation targets.
$5.00

FLUIDIZED-BED ENERGY TECHNOLOGY: Coming To A Boil (June 1978)
A report on the state-of-the-art of fluidized-bed combustion technology for cleaner and more efficient direct burning of coal. This report provides information on the companies now researching and developing fluidized-bed systems, the variety of possible large- and small- scale applications for the technology, its cost, and the barriers to widespread commercial use.
$45.00

CLEANING UP COAL: A Study of the Technology and Use of Cleaned Coal (March 1982)
A close examination of various coal-cleaning techniques, reporting on the state-of-operation and state-of-the-art of coal cleaning systems. The study examines the present status of sulfur- and ash-removing capabilities of preparation plants, costs and environmental effects of clean coal processes and utilities' experiences with burning cleaned coal. The study profiles experiences of 28 companies in four different areas connected with the technology: construction, ownership, operation, and use of this fuel. Reports also on the status of research and development of coal cleaning systems by the DOE, EPA and EPRI.

THE SCRUBBER STRATEGY: The How and Why of Flue Gas Desulfurization (April 1982)
An examination of the state-of-operation and state-of-the-art of flue gas desulfurization technology, including the various systems available; the problems of waste disposal and treatment methods; the SO_2 — reducing capabilities and reliability of systems in use; the power, water and reagents employed; operation and maintenance problems; marketing difficulties; and comparison of capital costs. The study also examines promising new experimental systems now being used.

Land Use

HOW TO JUDGE ENVIRONMENTAL PLANNING FOR SUBDIVISIONS: A Citizen's Guide (February 1981)
For the citizen on a community planning board, this guide provides the essential criteria for judging proposed land subdivisions. What are the right questions to pose to prevent poorly planned developments from becoming costly errors to the community? $3.95

THE INSIDER'S GUIDE TO OWNING LAND IN SUBDIVISIONS: How to Buy, Appraise and Get Rid of Your Lot (January 1981)
A manual for lot buyers. What to know about the environmental practices of a developer and the investment and residential value of a lot—before buying. Also the steps to take if you have already bought and find your lot not to be what you expected. $2.50

PROMISED LANDS 1: Subdivisions in Deserts and Mountains (October 1976)
A study of ten sites in the Southwest and West describing and evaluating the effects of the U.S. land subdivision industry operations on consumers and the environment. $20.00

PROMISED LANDS 2: Subdivisions in Florida's Wetlands (March 1977)
An analysis and evaluation of the environmental and consumer impact of nine Florida subdivisions. $20.00 (Check availability before ordering)

PROMISED LANDS 3: Subdivisions and the Law (January 1978)
An assessment of the effectiveness of land-sales and land-use laws in protecting the environment and consumers. $20.00

BUSINESS AND PRESERVATION: A Survey of Business Conservation of Buildings and Neighborhoods (May 1978)
An examination of corporate activities involving preservation and reuse of existing buildings and historic sites, and support of neighborhood redevelopment. $14.00

Occupational Safety and Health

AT WORK IN COPPER, Volumes 1, 2, 3 (April 1979)
A study of conditions at the 16 U.S. copper smelters, identifying worker safety and health risks, defining the best available worker protection techniques, and evaluating company, government and union efforts to protect employees. Set: $70.00

Volume 1 provides the findings of overall industry performance, with recommendations of feasible engineering controls. Explains criteria upon which smelter evaluations are based. $40.00

Volume 2 profiles smelters owned by ARCO, ASARCO, Cities Services, and Inspiration. $20.00

Volume 3 profiles smelters owned by Kennecott, Louisiana Land, Newmont Mining, and Phelps Dodge. $20.00

HOW OSHA ENFORCES THE LAW (January 1982)
The effectiveness of the Occupational Safety and Health Administration in protecting workers in the copper smelting industry. $15.00

Future Publications

ENERGY STORAGE
A study reporting on near-term applications of 5 energy storage technologies, all of which can help utilities to stretch the usefulness of existing power plants. They may postpone or eliminate the need for new generating capacity by storing off-peak electric power for use during hours of peak demand.

PROMISED LANDS 2 UPDATE
An up-to-the-present examination of land subdivision and sales practices in Florida. Some sites will be revisited, some recent sites will be examined for the first time.

INFORM Staff

Executive Director
Joanna Underwood

Director of Research
Robbin Blaine

Public Education Associate/Editor
Richard Allen

Director of Development
Joan Platt

Director of Administration
Viviane Arzoumanian

Project Directors
Mary Ann Baviello
Ruth Gallo
Cynthia Hutton
Kenneth Pollack
Sophie Weber

Research Associates
Robert Gould
Clifford Bob

Research Assistants
Alexandra Bowie
Caryn Halbrecht
Michael Jacobs

Public Education Associates
Risa Gerson
Susan Jakoplic

Copy Editor
Mary Ferguson

Administrative Assistants
Denise Ellis
Patricia Holmes
Patricia Jospe
Linda Post
Linda Vinecour

Student Interns
Michael Gergen
Ilene Green
Jonathan Kalb
Marybeth McCleery
Matthew Roberts
Lisa Rosenfield
Carol Steinsapir
Nancy Warren

Consultants
James Cannon, Energy
Howard Girsky, Publicity
Manuel Gomez, Occupational
Safety & Health
Perrin Stryker, Editorial
Daniel Wiener, Energy

Board of Directors

Timothy Hogen: Chairman
President
T.L. Hogen Associates

Susan Butler
Associate Director
Environmental Defense Fund

C. Howard Hardesty, Jr.
Partner
Corcoran, Hardesty, Ewart,
Whyte & Polito

Fred M. Hechinger
President
New York Times Company
Foundation

Lawrence S. Huntington
President
Fiduciary Trust Company
    Of New York

Martin Krasney
Consultant of Executive
    Development
ARCO

Kenneth F. Mountcastle, Jr.
Senior Vice President
Dean Witter Reynolds, Inc.

Barbara E. Niles

Kenneth Pollack
INFORM

Grant P. Thompson
Senior Research Associate
The Conservation Foundation

Edward Hallam Tuck
Partner
Shearman & Sterling

Joanna Underwood
INFORM

Frank A. Weil
Partner
Ginsberg, Feldman, Weil &
    Bress

Anthony Wolff
Journalist